Eleven Papers Translated from the Russian

AMERICAN MATHEMATICAL SOCIETY
TRANSLATIONS

Series 2

Volume 140

Eleven Papers Translated from the Russian

by

V. I. Bernik
N. A. Dmitriev
E. B. Dynkin
K.-H. Elster
L. A. Gutnik
R. Heine
R. I. Karpelevich
L. G. Kiseleva
A. F. Leont'ev
Yu. I. Manin
Yu. P. Razmyslov
L. A. Skornyakov
D. P. Skvortsov

AMERICAN MATHEMATICAL SOCIETY
PROVIDENCE, RHODE ISLAND

Translation edited by BEN SILVER

1980 *Mathematics Subject Classification* (1985 *Revision*). Primary 03C62, 11J69, 11J72, 14A20, 15A18, 15A48, 15A51, 16A38, 17B60, 18B10, 20M05, 30B50, 30B60, 32L25, 58A50, 90A05, 90C31; Secondary 03C10, 11M06, 20M30, 20M50, 26C10, 30C15, 68P20, 81E13.

Library of Congress Cataloging-in-Publication Data
Eleven papers translated from the Russian.
(American Mathematical Society translations, ISSN 0065-9290; ser. 2, v. 140)
Includes bibliographies.
1. Mathematics–1961– . I. Bernik, V. I. (Vasilii Ivanovich) II. Silver, Ben. III. Title: 11 papers translated from the Russian. IV. Series.
QA3.A572 ser. 2, vol. 140 510s [510] 88-26042 ISBN 0-8218-3116-X

Contents

Russian Contents*

*The American Mathematical Society scheme for transliteration of Cyrillic may be found at the end of index issues of *Mathematical Reviews*.

Amer. Math. Soc. Transl.
(2) Vol. **140**, 1988

On the Question of "How Many": Definition of the Notion of the Cardinality of Finite Sets in Some Arithmetical Systems

D. P. SKVORTSOV

In connection with the formal simulation of information retrieval systems [1] the problem of how to interpret query-and-reply predicates of various types arises. In particular, it is necessary to interpret the question "how many x have the property $\varphi(x)$?" (questions of this type are considered for example in [1] and [2]). In the interpretation of such a question the necessity of investigating the natural numbers and the operator defining the cardinality of finite sets naturally appears.

In this article we study the possibility of defining the cardinalities of finite sets of natural numbers in various formal languages for arithmetic (as one easily sees, this is equivalent to the definability of the operator $|\{x < y \colon \varphi(x)\}|$: "the cardinality of the set of all x less than y which satisfy the property $\varphi(x)$").(1) It appears that the smallest language in which such a definition is possible in the sense specified is the language of Presburger arithmetic [3]. At the same time there are extensions of the language of Presburger arithmetic in which the notion of cardinality is not definable. We can also find languages such that the notion of cardinality is definable both in these languages and all their extensions: for example, the language of Peano arithmetic ([4], Chapter 3, §1).

1980 *Mathematics Subject Classification* (1985 *Revision*). Primary 03C62; Secondary 03C10, 68P20.

Translation of Semiotika i Informatika, vyp. 24, VINITI, Moscow, 1985, pp. 41–57; MR **86m**: 03088.

(1)We remark that the concepts of *finitude* and *infinitude* for sets of natural numbers are definable in rather simple arithmetical languages:

$$\{x \colon \varphi(x)\} \text{ is finite } =\colon \exists y \forall x(\varphi(x) \supset x < y);$$

therefore the definition of the cardinality of *finite* sets is sufficient for a definition of the cardinality of arbitrary sets of natural numbers.

The basic notions and terminology are briefly sketched in the Introduction. However, for the notions of first-order statements and their proofs, a certain fluency in the treatment of these notions is desirable, presupposing an acquaintance with the elements of mathematical logic in Chapters 1 and 2 (and 3, in part) of Mendelson's book [4].

The author is indebted to V. K. Finn for posing the problem and for discussion of the results, and to A. L. Semenov for some useful conversations.

Introduction

By a *first-order language* we mean a sequence of *function symbols* ($f_i^{(n_i)}$) and *predicate symbols* ($P_i^{(n_i)}$) of various valencies $n_i > 0$ (in particular, 0-place function symbols are usually called *individual constants*). We suppose there is a fixed countable set Vr of *individual variables*. *Terms* and *formulas* of the first-order language L are defined in the usual way ([4], Chapter 2, §1), namely, according to the following inductive rules.

1. If $x \in \mathrm{Vr}$, then x is a term.

2. If $t_1, \ldots, t_{n_i}$ are terms, then $f_i^{(n_i)}(t_1, \ldots, t_{n_i})$ is a term and $P_i^{(n_i)}(t_1, \ldots, t_{n_i})$ is a formula (such a formula is called an *atomic formula* for the language L).

3. If φ and ψ are formulas, and $x \in \mathrm{Vr}$, then $(\varphi \mathbin{\&} \psi)$, $(\varphi \vee \psi)$, $(\varphi \supset \psi)$, $(\varphi \equiv \psi)$, $\neg\varphi$, $(\forall x)\varphi$, and $(\exists x)\varphi$ are formulas for the language L.

Individual variables will be denoted by the letters x, y, z, u, v, w, terms by t, r, s, and formulas by φ, ψ, η (with or without subscripts). For brevity we write $\varphi \in L$ (or $\varphi \in E(L)$) to mean that φ is a formula (respectively, an atomic formula) of the language L. *Free* and *bound* variables in formulas or terms are defined in the usual way ([4], Chapter 2, §1). We call the variables which occur freely in a formula or term its *parameters*; a formula or term is *closed* if it contains no parameters. When we write $\varphi(x_1, \ldots, x_n)$ or $t(x_1, \ldots, x_n)$ we mean that φ or t is a formula or term with the parameters $x_1, \ldots, x_n$. The *universal closure* of a formula $\varphi = \varphi(x_1, \ldots, x_n)$ denotes the formula $(\overline{\forall} \cdot \varphi) =: (\forall x_1) \cdots (\forall x_n)\varphi(x_1, \ldots, x_n)$. A formula φ of the language L is said to be *quantifier-free* if the quantifiers $\forall$ and $\exists$ do not occur in it; any quantifier-free formula can be put into *disjunctive normal form*, $\bigvee_i \mathbin{\&}_j \varphi_{ij}$, with each φ_{ij} either in $E(L)$ or of the form $\neg\varphi'_{ij}$ with φ'_{ij} in $E(L)$ ([4], Chapter 1, §3).

By a *model* M (for a first-order language L) with *universe* D (where D is a nonempty set) we mean a sequence of functions (operations) and predicates on D corresponding to the function and predicate symbols of the language L (with the appropriate valencies!). The model M determines the *values* of the terms and formulas of L in the usual way ([4], Chapter 2, §2): the value is defined by "substitution" of certain elements of D for the parameters in the term or formula; the value of a term is an element of D, and the value of a formula is its truth value (true or false). When speaking of a first-order language L we always suppose that some concrete model has been fixed for it, and the truth of a formula φ of L in that model is written $\models \varphi$. Two formulas φ and ψ are *equivalent* if $\models \overline{\forall} \cdot (\varphi \equiv \psi)$.

Let L_1 and L_2 be two first-order languages whose intended models have a common universe D. We say that the language L_1 is *contained in* L_2 ($L_1 \subseteq L_2$) if $(\forall\varphi)(\varphi \in L_1 \Rightarrow \varphi \in L_2)$. The language L_1 is *defined in* L_2 (denoted: $L_1 \underset{\sim}{\subset} L_2$) if for each $\varphi \in L_1$ there is a $\varphi' \in L_2$ such that $\models \overline{\forall} \cdot (\varphi \equiv \varphi')$; a definition of the language L_1 in L_2 is called *effective* if the function $(\varphi \mapsto \varphi')$ is algorithmically computable. We note that if a language L_1 is contained in L_2 then L_1 is certainly defined in L_2, even effectively. We say that the language L_2 is *strictly stronger than* L_1 (written $L_1 \underset{\not\sim}{\subset} L_2$) if $L_1 \underset{\sim}{\subset} L_2$ and $L_2 \underset{\sim}{\not\subset} L_1$. The languages L_1 and L_2 are of *equal strength* (or *interdefinable*, denoted $L_1 \sim L_2$) if $L_1 \underset{\sim}{\subset} L_2$ and $L_2 \underset{\sim}{\subset} L_1$. We remark that every first-order language L can be converted to a language L' of equal strength which does not contain function symbols ([4], Chapter 2, §9). We will sometimes exploit this possibility without explicit mention.

§1. Definability of the concept of the cardinality of finite sets

1.1. Let L be a first-order logic with equality (i.e., containing the binary predicate symbol $(x = y)$), for which we suppose the model with the universe $N = \{0, 1, 2, \dots\}$ has been fixed (this will be called the *standard model* for the language L on the set of natural numbers N). Here the standard model is assumed to be normal (in the terminology of [4], Chapter 2, §8); that is, the predicate symbol $(x = y)$ of L corresponds to the usual equality relation on N. By the *theory* $A(L)$ in the first-order language L we mean the set of all propositions of L which are true in the standard model.

We consider the *extended language* L^+ obtained from L by adding the following rule of construction.

If $\varphi(x, x_1, \dots, x_n)$ is a formula of the language L^+ then we introduce a function symbol $f_\varphi^{(n+1)}$, which we agree to describe as follows:

$$f_\varphi^{(n+1)} =: |\{x < y \colon \varphi(x, x_1, \dots, x_n)\}|;$$

or, equivalently, we introduce the atomic formula:

$$\rho_\varphi(x_1, \dots, x_n, y, z) =: (|\{x < y \colon \varphi(x, x_1, \dots, x_n)\}| = z).$$

(Warning: x is a bound variable in (1)!)

The truth of a formula of L^+ in the standard model is understood in a natural way, in accordance with the notation (1): the value of the term $f_\varphi^{(n+1)}(x_1, \dots, x_n, y)$ is the cardinality of the set $\{x < y \colon \models \varphi(x, x_1, \dots, x_n)\}$. As above, the *theory* $A(L^+)$ denotes the set of all true propositions in L^+.

We also consider an auxiliary *intermediate language* L' in which the function symbol $f_\varphi^{(n+1)}$ is added only for formulas φ of the initial language L. Here $L \subseteq L' \subseteq L^+$.

We will say that the concept of the cardinality of finite sets is *definable in* L if $L^+ \underset{\sim}{\subset} L$ (and then clearly $L^+ \sim L$).

It is easy to check that:

LEMMA 1.1. $L^+ \underset{\sim}{\subseteq} L \Leftrightarrow L' \underset{\sim}{\subseteq} L$.

Thus, in order that the concept of the cardinality of a finite set should be definable in the language L, it is necessary and sufficient that for each $\varphi(x, x_1, \dots, x_n) \in L$ there should be some $\varphi'(x_1, \dots, x_n, y, z) \in L$ such that

$$\models \overline{\forall} \cdot [\varphi'(x_1, \dots, x_n, y, z) \\ \equiv (|\{x < y \colon \varphi(x, x_1, \dots, x_n)\}| = z)]. \tag{2}$$

REMARK. In this connection, in order that L^+ should be *effectively defined* in L it is necessary and sufficient that the function $(\varphi \mapsto \varphi')$ satisfying (2) should be algorithmically computable.

1.2. Supposing that the language L contains the constant 0 and the successor function $'$ $(x' = x + 1)$, we consider the elementary theory (in the sense of [4], Chapter 2, §3) in the language L^+ with the following nonlogical axioms (where φ is an arbitrary formula of L^+):(2)

(a) all formulas in $A(L)$;
(b) $(\forall x_1) \cdots (\forall x_n)(\forall y)(\exists_1 z)(|\{x < y \colon \varphi(x, x_1, \dots, x_n)\}| = z)$;
(c0) $(|\{x < 0 \colon \varphi(x, x_1, \dots, x_n)\}| = 0)$;
(c1) $\varphi(y, x_1, \dots, x_n) \supset (|\{x < y' \colon \varphi(x, x_1, \dots, x_n)\}| = (|\{x < y \colon \varphi(x, x_1, \dots, x_n)\}|)')$;
(c2) $\neg\varphi(y, x_1, \dots, x_n) \supset (|\{x < y' \colon \varphi(x, x_1, \dots, x_n)\}| = |\{x < y \colon \varphi(x, x_1, \dots, x_n)\}|)$;
(d) the *induction scheme*: $\varphi(0) \mathbin{\&} (\forall x)(\varphi(x) \supset \varphi(x')) \supset (\forall x)\varphi(x)$.

Derivability of a formula of the language L^+ in this elementary theory is understood in the usual way ([4], Chapter 2, §3); we denote by $A^+(L)$ the set of all its theorems (derivable formulas). It is easy to show that $A^+(L) \subseteq A(L^+)$. Furthermore, the *axioms of equality* for L^+ are derivable in the theory $A^+(L)$, i.e., all formulas of the form

$$(x = y) \supset (\varphi(x) \equiv \varphi(y)) \tag{3}$$

for all $\varphi \in L^+$, and also the *axioms of substitution equivalence in "cardinality terms"*:

$$(\forall x)(\varphi(x, x_1, \dots, x_n) \equiv \psi(x, x_1, \dots, x_n)) \\ \supset (\forall y)(|\{x < y \colon \varphi(x, x_1, \dots, x_n)\}| \\ = |\{x < y \colon \psi(x, x_1, \dots, x_n)\}|). \tag{4}$$

1.3. We take up the question as to when $A^+(L) = A(L^+)$; that is, when is the elementary theory $A^+(L)$ an axiomatization for the semantically determined theory $A(L^+)$? It turns out that this question is intimately connected with the question of the relative strengths and complexities of the languages L and L^+, and also of the theories $A(L)$ and $A(L^+)$.

(2)The notation $(\exists_1 z)\varphi(z)$ (there is one and only one z) is introduced in [4], Chapter 2, §8, where E appears in place of $\exists$.

THEOREM 1.1. *If* $A^+(L) = A(L^+)$, *then* ($A(L)$ *is decidable* $\Leftrightarrow$ $A(L^+)$ *is decidable*).

PROOF. If the theory $A(L)$ is decidable, then $A^+(L)$ is recursively axiomatizable, and $A(L^+)$ is complete. But, as is well known, a complete recursively axiomatizable theory is decidable ([4], Chapter 3, §6, Exercise 3.47(b)).

THEOREM 1.2. *If* $L^+ \underset{\sim}{\subset} L$ *then* $A^+(L) = A(L^+)$.

The proof uses the following lemma.

LEMMA 1.2. *If* $\psi \in E(L')$ *and* $\psi' \in E(L)$, *with* $\models \overline{\forall}\cdot(\psi \equiv \psi')$, *then* $A^+(L) \vdash (\psi \equiv \psi')$.

The proof is based on the application of the induction scheme (c) and the fact that if $\psi =: (|\{x < y \colon \varphi(x, x_1, \ldots, x_n)\}| = z)$, and $\models (\psi \equiv \psi')$, then the analogues of axioms (b), (c0), (c1), and (c2) with the substitution of ψ for ψ' are true formulas of the language L; that is, they are derivable in the theory $A(L)$.

COROLLARY. *If* $L^+ \underset{\sim}{\subset} L$, *then for all* $\varphi \in L^+$ *there is a* $\varphi' \in L$ *such that* $A^+(L) \vdash (\varphi \equiv \varphi')$.

The proof proceeds by induction on the structure of the formula φ of the language L^+, where for the proof of $(\varphi \equiv \varphi')$ the derivability in $A^+(L)$ of formulas of the forms (3) and (4) is used.

Now we prove Theorem 1.2. Let $L^+ \underset{\sim}{\subset} L$ and $\varphi \in A(L^+)$; that is, $\varphi \in L^+$ and $\models \varphi$. We have some $\varphi' \in L$ with $A^+(L) \vdash (\varphi \equiv \varphi')$ (that is, $\models (\varphi \equiv \varphi')$), and since $\models \varphi$, we have $\models \varphi'$. Since $\varphi' \in L$, we have $\varphi' \in A(L)$, so $A^+(L) \vdash \varphi'$, and this means that $A^+(L) \vdash \varphi$, which is what was to be shown.

COROLLARY. *If* $A(L)$ *is decidable and the language* L^+ *is defined in* L, *then* L^+ *is effectively defined in* L *and* $A(L^+)$ *is also decidable.*

In the following two sections we consider as examples certain arithmetical first-order languages L, and we will describe the associated languages L' and L^+, as well as the theories $A(L^+)$ and $A^+(L)$. There we will show in which of them the concept of the cardinality of a finite set can or cannot be defined, and also we will illustrate how the concepts "$L^+ \underset{\sim}{\supset} L$" and "$A^+(L) = A(L^+)$" are interrelated, and show that the decidability of the theories $A(L)$ and $A^+(L)$ may occur in concrete situations.

§2. The language of Presburger arithmetic and some of its fragments

We consider the following three first-order languages (whose standard models on the set of natural numbers N are defined in a natural way).

1. The *minimal arithmetical language* L_0 contains the constant 0 and the successor function $'$.

2. The *language of ordered arithmetic* L_V contains 0, $'$, and the predicate $(x < y)$.

3. The *language of Presburger arithmetic* L_P contains 0, $'$, and the function $+$ (addition).

Here $L_O \subseteq L_V \underset{\sim}{\subseteq} L_P$, since $(x < y) \equiv (\exists z)(x + z' = y)$. For every natural number n we define the *numeral* (as a term in the language L_O) $\overline{n} = 0^{'\cdots'}$ (n times), and we also define the term in the language L_P, $\overline{n} \cdot x =: (x + \cdots + x)$ (n times). We consider the auxiliary language $\tilde{L}_P$ (interdefinable with L_P) obtained by adjoining the predicate $(x < y)$ and the predicates of *number-theoretic congruence* $(x \equiv y \pmod{\overline{n}})$ for all n to L_P (where $(x \equiv y \pmod{\overline{n}}) \equiv (\exists z)((x = y + \overline{n} \cdot z) \vee (y = x + \overline{n} \cdot z)))$.

LEMMA 2.1 [3]. *For all* $\varphi(x_1, \ldots, x_n) \in L_P$ *there is a quantifier-free* $\varphi'(x_1, \ldots, x_n) \in \tilde{L}_P$ *such that* $\models (\varphi \equiv \varphi')$.

The proof is presented in [4] (Chapter 3, §1, Exercise 3.7D) or in detail in [5] (Chapter VII, §4.3). Hence we have

THEOREM 2.1. *Presburger arithmetic* $A(L_P)$ *is decidable.*

COROLLARY. *The theories* $A(L_O)$ *and* $A(L_V)$ *are decidable.*

2.1. THEOREM 2.2. $L_P^+ \underset{\sim}{\subseteq} L_P$. (*The notion of the cardinality of a finite set is definable in Presburger arithmetic and in fact effectively.*)

The proof is based on the following lemma ([5], Chapter VII, §4.3).

LEMMA 2.2. *Any quantifier-free* $\varphi(x, x_1, \ldots, x_k) \in \tilde{L}_P$ *is equivalent to a quantifier-free* $\varphi'(x, x_1, \ldots, x_k) \in \tilde{L}_P$ *of the following form:*

$$\varphi' =: \bigvee_i [\varphi_i'(x, x_1, \ldots, x_k) \;\&\; \varphi_i''(x_1, \ldots, x_k)],$$

where

$$\varphi_i' =: [(x \equiv \overline{m}_i (\operatorname{mod} \overline{n}_i)) \;\&\; (t_i < \overline{l}_i x + s_i) \;\&\; (\overline{l}_i x + s_i < r_i)], \tag{5}$$

m_i, n_i, $l_i \in N$ ($m_i < n_i$) *and* t_i, r_i, s_i *are terms of* L_P, *not containing the variable* x.

In order to prove Theorem 2.2 it is sufficient in view of Lemma 1.1 to associate with each $\varphi(x, x_1, \ldots, x_k) \in L_P$ a formula $\psi(x_1, \ldots, x_k, y, z)$ such that

$$\models \overline{\forall} \cdot [\psi(x_1, \ldots, x_k, y, z) \equiv (|\{x < y \colon \varphi(x, x_1, \ldots, x_k)\}| = z)]. \tag{6}$$

Here, in view of Lemma 2.2, it is enough to consider only quantifier-free formulas φ, and since

$$|\{x < y \colon (\varphi_1 \vee \varphi_2)\}| = |\{x < y \colon \varphi_1\}| + |\{x < y \colon \varphi_2\}| - |\{x < y \colon (\varphi_1 \;\&\; \varphi_2)\}|, \tag{7}$$

we may limit ourselves to the consideration of formulas $\varphi =: \varphi_i'(x, x_1, \ldots, x_k)$ of the form (5). It is then easy to see that

$$\models \overline{\forall} \cdot \left[(|\{x < y \colon \varphi_i'(x, x_1, \ldots, x_k)\}| = z) \equiv \left[\neg \left(\bigvee_{p=0}^{l_i n_i} \psi_p \right) \& (z = 0) \right] \bigvee_{p=0}^{l_i n_i} (\psi_p \;\&\; \eta_p) \right],$$

where

$$\psi_0 =: (t_i < s_i + \overline{(l_i \cdot m_i)}) \ \& \ (s_i + \overline{(l_i \cdot m_i)} < r_i),$$

$$\eta_0 =: ([(\overline{(l_i \cdot n_i)} \cdot z + s_i + \overline{(l_i \cdot n_i + 1)} > r_i) \\ \vee (\overline{n}_i \cdot z + \overline{(m_i + 1)} > y)] \ \& \ ((\overline{(l_i \cdot n_i)} \cdot z + s_i + \overline{(l_i n_i)} \\ < r_i + \overline{(l_i n_i)}) \ \& \ (\overline{n}_i \cdot z + \overline{m}_i < y + \overline{n}_i)),$$

and for $p = 1, \dots, l_i n_i$,

$$\psi_p =: (s_i + \overline{(l_i \cdot m_i)} < t_i') \ \& \ (t_i + \overline{p} < r_i) \ \& \ (t_i + \overline{p} \equiv s_i + \overline{(l_i m_i)} (\operatorname{mod} \overline{(l_i n_i)})),$$

$$\eta_p =: ([(\overline{(l_i \cdot n_i)} \cdot z + t_i + \overline{(p+1)} > r_i) \vee (\overline{(l_i \cdot n_i)} \cdot z + t_i + \overline{(p+1)} > s_i + \overline{l}_i \cdot y)] \\ \& \ (\overline{(l_i \cdot n_i)} \cdot z + t_i + \overline{p} < r_i + \overline{(l_i \cdot n_i)}) \\ \& \ (\overline{(l_i \cdot n_i)} \cdot z + t_i + \overline{p} < s_i + \overline{l}_i y + \overline{(l_i n_i)})).$$

The theorem is proved.

COROLLARY.

1. $A^+(L_p) = A(L_p^+)$.
2. *The theory $A(L_p^+)$ is decidable.*

2.2. THEOREM 2.3.

1. $L_V' \sim L_V^+ \sim L_P$.
2. $A^+(L_V) = A(L_V^+) = A(L_P)$.

PROOF. 1. $L_P \lesssim L_V'$, since

$$(x + y = z) \equiv (x \leq z) \ \& \ (|\{u < z \colon (x \leq u)\}| = y), \tag{8}$$

where

$$(v \leq w) \equiv (v < w) \vee (v = w).$$

For the proof of 2 it is sufficient to derive the Peano axioms for addition in $A^+(L_V)$ ([4], Chapter 3, §1):

$$(x + 0 = x), \tag{9}$$

$$(x + y = z) \supset (x + y' = z'), \tag{10}$$

$$(\forall x)(\forall y)(\exists_1 z)(x + y = z). \tag{11}$$

Here (10) and (11) are proved by induction on y with the use of (c1), while (9) follows from

$$(y \leq x) \supset (|\{u < y \colon (x \leq u)\}| = 0),$$

which is proved by induction on y using (c2).

COROLLARY. *If $L_V \lesssim L \lesssim L_P$, then $L' \sim L^+ \sim L_P$ and $A^+(L) = A(L^+) = A(L_P)$.*

REMARK. In order to see the inclusion of Theorem 2.3, one also shows that the language L_V is weaker than L_P; that is, $L_P \not\lesssim L_V$. Indeed, in [5] (Chapter V, §3) it is shown that every $\varphi \in L_V$ is equivalent to some quantifier-free $\varphi' \in L_V$. Hence for every formula with *one* parameter $\varphi(x)$ in the language L_V, one of $\{n \in N \colon \models \varphi(n)\}$ and $\{n \in N \colon \models \neg\varphi(n)\}$ is finite, and therefore the predicate $(x \equiv \overline{m} \ (\operatorname{mod} \overline{n}))$ is not definable in L_V for $n > 1$.

2.3. We now consider the minimal language L_0. The following lemma can be proved in a way similar to 2.1.

LEMMA 2.3. *Every $\varphi \in L_0$ is equivalent to a quantifier-free $\varphi' \in L_0$.*

Hence it is easy to conclude that the predicate $(x < y)$ is not definable in L_0, and accordingly that $L_0 \subsetneq L_V$.

THEOREM 2.4.
1. $L'_0 \sim L_V$.
2. $L_0^+ \sim L_P$.

(Thus there is a language L, namely L_0, with $L \subsetneq L' \subsetneq L^+$.)

PROOF. The second claim follows from the first in view of Theorem 2.3. In the first claim, $L_V \precsim L'_0$ since

$$(x < y) \equiv \neg(|\{z < y \colon (z = x)\}| = 0). \tag{12}$$

In order to establish that $L'_0 \subseteq L_V$, in view of Lemma 2.3 it is sufficient to establish the following lemma.

LEMMA 2.4. *For each quantifier-free formula $\varphi(x, x_1, \ldots, x_n) =: \bigvee_{i=1}^{\omega} \varphi_i$, where*

$$\varphi_i =: \left(\mathop{\&}_j \varphi'_{ij} \;\&\; \mathop{\&}_k \varphi''_{ik} \right), \tag{13}$$

and φ'_{ij} and φ''_{ik} are formulas of the forms $(u = \overline{m})$, $\neg(u = \overline{m})$, $(u = v + \overline{m})$, or $\neg(u = v + \overline{m})$, with $u, v \in \{x, x_1, \ldots, x_n\}$, where φ'_{ij} contains x and φ''_{ik} does not, there is a formula $\varphi'(x_1, \ldots, x_n, z) \in L_V$ such that

$$\vDash \overline{\forall} \cdot [\varphi'(x_1, \ldots, x_n, y, z) \equiv (|\{x < y \colon \varphi(x, x_1, \ldots, x_n)\}| = z)]. \tag{14}$$

We show by induction on ω that there is such a formula φ' having the form

$$\varphi' =: \bigvee_j [\varphi'_j(x_1, \ldots, x_n, y) \;\&\; (z = t_j)], \tag{15}$$

where t_j is the term $\overline{m}_j$, or $(y - \overline{m}_j)$, (where the notation $[z = y - \overline{m}_j]$ means naturally $[y = z + \overline{m}_j]$, i.e., $[y = z^{\prime \cdots \prime} (m_j \text{ primes})]$).

Basis of the induction: $\omega = 1$.

Case I. For some j, φ'_{ij} is $(x = t)$, where $t = \overline{m}$ or $t = (x_1 + \overline{m})$.

Here we set $\varphi' =: [\neg\tilde{\varphi} \;\&\; (z = 0)] \vee [\tilde{\varphi} \;\&\; (z = \overline{1})]$, where $\tilde{\varphi}$ is

$$\tilde{\varphi} =: \left[\mathop{\&}_j \tilde{\varphi}'_{1j} \;\&\; \mathop{\&}_k \varphi''_{1k} \;\&\; (t < y) \right],$$

and $\tilde{\varphi}'_{1j}$ is the result of substituting the term t for x in φ'_{ij}.

Case II. For some j, φ'_{1j} is $(x_1 = x + \overline{m})$.

This is handled like the preceding case, except that the substitution of the term $t =: (x_1 - \overline{m})$ for x is carried out (in the spirit of (15)).

Case III.

$$\left(\mathop{\&}_j \varphi'_{1j} \right) =: \left[\mathop{\&}_{a=1}^{d} \neg(x = \overline{m}_a) \;\&\; \mathop{\&}_{b=1}^{e} \neg(x = y_b + \overline{m}_b) \;\&\; \mathop{\&}_{c=1}^{f} \neg(y_c = x + \overline{m}_c) \right];$$

where $y_b,\ y_c \in \{x_1, \ldots, x_n\}$. For $g \leq h = (d+e+f)$ we define a formula $\psi_g(x_1, \ldots, x_n, y)$ so that for all $x_1, \ldots, x_n$ and y in N

$$\begin{aligned}\vDash \psi_g(x_1, \ldots, x_n, y) \Leftrightarrow (|(\{\overline{m}_a \colon a\} \cup \{(y_b + \overline{m}_b) \colon b\} \\ \cup \{(y_c - \overline{m}_c) \geq 0 \colon c\}) \cap \{z \colon z < y\}| = g),\end{aligned}$$

and we put

$$\varphi' =: \left(\left[\neg \left(\mathop{\&}_{k} \varphi''_{1k} \right) \ \& \ (z = 0) \right] \vee \bigvee_{g=0}^{h} \left[\left(\mathop{\&}_{k} \varphi''_{1k} \ \& \ \psi_g \right) \ \& \ (z + \overline{g} = y) \right] \right).$$

Inductive step. Let $\varphi =: (\bigvee_{i=0}^{\omega} \varphi'_i) = (\varphi'_0 \vee \psi)$, where $\psi =: (\bigvee_{i=1}^{\omega} \varphi'_i)$. In view of (7),

$$|\{x < y \colon \varphi\}| = |\{x < y \colon \varphi'_0\}| + |\{x < y \colon \psi\}| - |\{x < y \colon \eta\}|,$$

where $\eta =: (\bigvee_{i=1}^{\omega} (\varphi'_0 \ \& \ \varphi'_i))$. By the induction hypothesis,

$$\vDash \overline{\forall} \cdot [(|\{x < y \colon \varphi'_0\}| = z) \equiv \bigvee_{j} (\varphi'_j \ \& \ (z = t_j))],$$

$$\vDash \overline{\forall} \cdot [(|\{x < y \colon \psi\}| = z) \equiv \bigvee_{k} (\psi'_k \ \& \ (z = t_k))],$$

$$\vDash \overline{\forall} \cdot [(|\{x < y \colon \eta\}| = z) \equiv \bigvee_{l} (\eta'_l \ \& \ (z = t_l))],$$

i.e.,

$$\vDash \tilde{\forall} \cdot [(|\{x < y \colon \varphi\}| = z) \equiv \bigvee_{j,k,l} (\varphi'_j \ \& \ \psi'_k \ \& \ \eta'_l \ \& \ [z = t_j + t_k - t_l])]. \tag{16}$$

In order to obtain the desired formula φ' we need to replace the expression $[z = (t_j + t_k - t_l)]$ in each of the disjunctive factors of (16) (apart from those which are identically false and may therefore be omitted) by $[z = t]$, where t is a term of the form (15) (if it is not already in this form, of course). For this, it is sufficient to remark that

$$\vDash [(\varphi'_j \ \& \ \psi'_k \ \& \ \eta'_l) \supset (0 \leq (t_j + t_k - t_l) \leq y)],$$

and hence if, say, $t_j = \overline{m}_j$, $t_k = \overline{m}_k$, and $t_l = (y - \overline{m}_l)$, then

$$\vDash [(\varphi'_j \ \& \ \psi'_k \ \& \ \eta'_l) \supset y \leq (\overline{m}_j + \overline{m}_k + \overline{m}_l)],$$

i.e.,

$$\begin{aligned}&\vDash \overline{\forall} \cdot (\varphi'_j \ \& \ \psi'_k \ \& \ \eta'_l \ \& \ [z = t_j + t_k - t_l]) \\ &\equiv \bigvee_{m \leq (m_j + m_k + m_l)} (\varphi'_j \ \& \ \psi'_k \ \& \ \eta'_l \ \& \ (y = \overline{m}) \ \& \ z = (\overline{m}_j + \overline{m}_k + \overline{m}_l - \overline{m})).\end{aligned}$$

The remaining possibilities for the form of the terms t_j, t_k, and t_l are treated similarly. The theorem is proved.

Clearly the technical difficulty of the proof is connected with the necessity to exploit relation (7), without having addition at our disposal in the language L_V.

COROLLARY 1. *If* $L_0 \underset{\sim}{\subseteq} L \underset{\sim}{\subseteq} L_P$, *then* $L^+ \sim L_P$.

COROLLARY 2. *If* $L_0 \underset{\sim}{\subseteq} L$ *and* $L^+ \underset{\sim}{\subseteq} L$, *then* $L_P \underset{\sim}{\subseteq} L$.

Thus if we restrict ourselves to only those languages which contain L_0, then among these *the language of Presburger arithmetic* L_P *turns out to be the smallest language admitting a definition of the concept of the cardinality of a finite set.* Now one would like to show that any language containing L_P also admits a definition of the concept of the cardinality of a finite set. However, this turns out not to be the case. This will be established in the next section, but for the present we will study the connection between $A^+(L_0)$ and $A(L_0^+)$.

2.4. LEMMA 2.5. *In the theory* $A^+(L_0)$ *the following formula is derivable:*

$$(\forall x)(|\{z < x \colon (z = x)\}| = 0). \tag{17}$$

PROOF (carried out in the theory $A^+(L_0)$). We suppose that, for some y, $\neg(|\{z < y \colon (z = y)\}| = 0)$. Now we have

$$(\forall x)((|\{z < x \colon (z = y)\}| = 0) \supset \neg(x = y)),$$

that is, in view of (c0) and (c2),

$$(|\{z < 0 \colon (z = y)\}| = 0)$$

and

$$(|\{z < x \colon (z = y)\}| = 0) \supset (|\{z < x' \colon (z = y)\}| = 0).$$

Hence by induction (axiom (d)) we conclude that $(|\{z < x \colon (z = y)\}| = 0)$ for all x. Now taking $x = y$ we arrive at a contradiction to our previous assumption. The lemma is proved.

THEOREM 2.5. $A^+(L_0) = A(L_0^+) = A(L_P)$.

PROOF. It is enough to derive the Peano axioms for addition (9)–(11) in $A^+(L_0)$. For this it is sufficient, using the definition (12), to derive those properties of the predicate $(x < y)$ which were used in the derivation of (9)–(11) in the proof of Theorem 2.3. As is easily shown (in detail in [5], Chapter VI, §5.2), these are

$$(x < x'), \tag{18}$$

$$(x < y) \;\&\; (y < z) \supset (x < z), \tag{19}$$

$$\neg(x < x). \tag{20}$$

Here (18) follows from $(x = x)$ in view of (c1); (19) is demonstrated by induction using (c1) and (c2); and (20) is a reformulation of (17). The theorem is proved.

§3. Some extensions of the language of Presburger arithmetic

We will now consider arithmetical languages containing the language L_P, and we will show that there are such languages for which there is no definition of the concept of cardinality for finite sets.

3.1. *The language of recursive arithmetic* L_R. This contains a function symbol $\cdot$ for multiplication in addition to the symbols 0, $'$, and $+$. In this language every recursive (algorithmically decidable) predicate on N is also definable, and its theory $A(L_R)$ is undecidable ([4], Chapter 3, §§3, 6).

THEOREM 3.1. *If* $L_R \subsim L$ *then* $L^+ \subsim L$.

PROOF. In view of Lemma 1.1, it is sufficient to associate with each $\varphi(x, x_1, \ldots, x_n) \in L$ a formula $\varphi'(x_1, \ldots, x_n, y, z) \in L$ satisfying (2). For this we set

$$\begin{aligned}\varphi'(x_1, \ldots, x_n, y, z) =:\ & (\exists u)[(\mathrm{lh}(u) = z) \\ & \& \ (\forall v)[(v < z) \supset (0 < (u)_v \leq y) \ \& \ \varphi((u)_v =: 1, x_1, \ldots, x_n)] \\ & \& \ (\forall x)[(x < y) \ \& \ \varphi(x, x_1, \ldots, x_n)) \supset (\exists v)((v < z) \ \& \ ((u)_v = x + 1))] \\ & \& \ (\forall v)(\forall w)[(v < z) \ \& \ (w < z) \ \& \ \neg(v = w) \supset \neg((u)_v = (u)_w)]]. \end{aligned} \tag{21}$$

Here the definability in L_R of the following primitive recursive functions is used ([4], Chapter 3, §2).

$\mathrm{lh}(u)$ is the number of nonzero exponents occurring in the decomposition of the number u as a product of powers of primes.

$(u)_v$ is the exponent of the vth prime number in the decomposition of the number u as a product of powers of primes.

$$(w \dot{-} 1) = (w - 1) \text{ if } w \neq 0, \text{ and } (w \dot{-} 1) = 0 \text{ if } w = 0.$$

The theorem is proved.

Thus *in the language* L_R *and any of its extensions the notion of the cardinality of finite sets is definable, and effectively so.* We will now indicate some languages lying between L_P and L_R in which the concept of cardinality is not definable.

3.2. We consider the following two predicates on N:

$$d_2(x) \equiv (\exists z)(x = 2^z), \tag{22}$$

$$d_2(x, y) \equiv (\exists z)(x = 2^z) \ \& \ (\exists u)(y = x \cdot u), \tag{23}$$

and also the languages L_{Q_0} and L_Q obtained by adjoining the predicates $d_2(x)$ or $d_2(x, y)$ respectively to L_P. (*Remark.* Here we could take any natural number $n > 1$ as the base for the exponentiation in (22) or (23) in place of the base 2, and all the results would generalize readily to that context.) Here we have $L_P \subseteq L_{Q_0} \subsim L_Q$, since $\models (\forall x)[d_2(x) \equiv d_2(x, x)]$.

THEOREM 3.2. *The theory* $A(L_Q)$ *is decidable.*

The proof is based on an embedding of $A(L_Q)$ into Büchi's decidable arithmetic [6]; we will not dwell on the technical details.

COROLLARY. *The theory* $A(L_{Q_0})$ *is decidable.*

THEOREM 3.3. $L'_Q \sim L^+_Q \sim L_R$. *That is, the adjunction to the language* L_Q *of the notion of the cardinality of finite sets leads to a definition of every recursive function.*

For the proof we consider an auxiliary language L_E, obtained by adjoining the function 2^x to the language L_P.

LEMMA 3.1. $(L_Q \cup L_E) \sim L_R$.

PROOF. In view of [7], §3, it is sufficient to define the divisibility predicate $x|y \equiv (\exists z)(x \cdot z = y)$ in $(L_Q \cup L_E)$. This we do as follows.

1. In L_E we define the predicate $(\mathrm{lh}(x) = y) \equiv$ "y is the length of the number $x \neq 0$ in binary notation" by

$$(\mathrm{lh}(x) = y) \equiv \neg(x = 0) \ \&\ (x < 2^y) \ \&\ (\forall z)(z < y \supset 2^z \leq x).$$

2. We arrange to enumerate the digits in the binary notation for x in ascending order, where we begin with 0; the ith digit in the binary expansion of x will be denoted $[x]_i$ (so $x = \sum_{i<\mathrm{lh}(x)}([x]_i \cdot 2^i)$, where $[x]_i \in \{0, 1\}$). We now define the following predicates in $(L_E \cup L_Q)$:

$$([x]_y = 1) \equiv \neg([x]_y = 0);$$

$$\begin{aligned}(x|y) \equiv (\exists z)[([z]_0 = 1) \ &\&\ ([z]_y = 1) \ \&\ (\mathrm{lh}(z) = y + 1) \\ &\&\ (\forall u)(\forall v)([(u < v < \mathrm{lh}(z)) \ \&\ ([z]_u = 1) \ \&\ ([z]_v = 1) \\ &\&\ (\forall w)(u < w < v \supset ([z]_w = 0))] \supset (v = u + x))].\end{aligned} \tag{24}$$

The lemma is proved.

LEMMA 3.2. $L_E \underset{\sim}{\subsetneq} L'_{Q_0}$.

Indeed,

$$(x = 2^y) \equiv [d_2(x) \ \&\ (|\{z < x \colon d_2(z)\}| = y)]. \tag{25}$$

Thus $L_R \underset{\sim}{\subsetneq} L'_{Q_0}$. The theorem is proved.

REMARK. Here $L_{Q_0} \underset{\not\sim}{\subsetneq} L_E$ and $L_Q \underset{\not\sim}{\subsetneq} L_R$, since the predicate $(x = 2^y)$ is not definable in L_Q. (This may be shown by using the embedding of L_Q into Büchi's arithmetic.) Thus $L_{Q_0} \underset{\not\sim}{\subsetneq} L'_{Q_0}$ and $L_Q \underset{\sim}{\subsetneq} L'_Q$; that is, *in the languages L_Q and L_{Q_0}, the notion of the cardinality of finite sets is not definable.*

OPEN PROBLEM. *Describe the languages L'_{Q_0} and $L^+_{Q_0}$.*

3.4. In view of Theorems 3.2 and 3.3, we have a language $L = L_Q$ for which the theory $A(L)$ is decidable while $A(L^+)$ is not; that is, *by adding a definition of the cardinality of finite sets we lose the decidability of the theory in this case.* In view of Theorem 1.1, this means also that $A^+(L_Q) \neq A(L^+_Q)$. We present a description of the theory $A^+(L_Q)$.

We consider the transformation which associates to each $\varphi \in L^+_Q$ an equivalent $\tilde{\varphi} \in L_R$ as follows. First we construct from φ a formula $\varphi^* \in L^+_R$ by changing $d_2(x, y)$ into its representation according to (23), and then we use the definition of L^+_R in L_R, described in Theorem 3.1.

THEOREM 3.4. *If $\varphi \in L^+_Q$ then $A^+(L_Q) \vdash \varphi$ iff $PA \vdash \tilde{\varphi}$, where PA is the set of axioms for Peano arithmetic* (*the system S in the terminology of* [4], *Chapter* 3, §1).

We will outline the proof in its most general features, omitting the technical details.

$\Rightarrow$. The proof proceeds by induction on the derivation of φ in the theory $A^+(L_Q)$, using the following two lemmas.

LEMMA 3.3. *If* $\varphi \in L_Q$ *and* $\models \varphi$, *then* $PA \vdash \varphi^*$ (*the system* PA *is "complete for the language* L_Q*"*).

The proof is based on the formalization within PA of the embedding of the language L_Q into Büchi's arithmetic [6].

LEMMA 3.4. *If* φ *is an axiom of* $A^+(L_Q)$ *of groups* (b), (c), *or* (d), *then* $PA \vdash \tilde{\varphi}$.

Indeed, the axioms of group (d) are transformed into induction axioms of PA, and for the axioms of groups (b) and (c) the proof is based on the formalization in PA of the corresponding properties of the formulas (21).

$\Leftarrow$. We use the inverse transformation associating to each $\psi \in L_R$ a formula $\psi' \in L_Q^+$, obtained from ψ by converting atomic subformulas of the form $(x{\cdot}y = z)$ into their definitions in L_Q^+ (according to Lemmas 3.1 and 3.2).

LEMMA 3.5. *If* $PA \vdash \psi$, *then* $A^+(L_Q) \vdash \psi'$.

In order to see this, one derives the transforms of the Peano axioms for multiplication in $A^+(L_Q)$. Their proof requires the verification of the derivability of certain properties of the relation $(x|y)$ defined according to (24), and this in turn depends on the verification of properties of the relation $(x = 2^y)$, defined in (25). Briefly speaking, the complete verification becomes a rather cumbersome routine exercise.

LEMMA 3.6. *If* $\varphi \in L_Q^+$, *then* $A^+(L_Q) \vdash [\varphi \equiv (\varphi^*)']$.

The proof depends on the verification of the derivability of the formula $[d_2(x,y) \equiv ([d_2(x,y)]^*)']$, which proceeds by induction on x and y.

LEMMA 3.7. *If* $\varphi \in L_Q^+$, *then* $A^+(L_Q) \vdash [(\varphi^*)' \equiv (\tilde{\varphi})']$.

The proof is based on the fact that $A^+(L_Q) \vdash (\tilde{\varphi})'$ if φ is an axiom of the system $A^+(L_Q)$ in groups (b) and (c), which follows from Lemmas 3.5 and 3.4.

COROLLARY. *If* $\varphi \in L_Q^+$ *then* $A^+(L_Q) \vdash [\varphi \equiv (\tilde{\varphi})']$.

Thus in view of Lemma 3.5, we conclude that PA $\vdash \tilde{\varphi} \Rightarrow A^+(L_Q) \vdash \varphi$. The theorem is proved.

3.5. From Theorem 3.3 it follows that:

COROLLARY.
1. *If* $L_Q \subsetneqq L \subsetneqq L_R$, *then* $L' \sim L^+ \sim L_R$.
2. *If* $L_Q \subsetneqq L$, *then* $L' \sim L^+ \sim (L \cup L_R)$.

Thus if $L_Q \subsetneqq L$ then $(L^+ \subsetneqq L)$ iff $(L_R \subsetneqq L)$. In other words, *among the extensions of the language* L_Q, *all extensions of* L_R, *and only those, admit definitions of the notion of the cardinality of finite sets.* The question remains open, *whether there is any arithmetical language*—other than L_P and extensions of L_R—*admitting a definition of cardinality*; however, such a language may certainly only be sought among the languages which contain L_P without containing L_Q.

REFERENCES

1. V. K. Finn, *Formal definition of the concept of an information-retrieval system*, Nauchn.-Tekhn. Informatsiya (VINITI) Ser. 2: Informatsionnye Protsessy i Sistemy **1981**, no. 5, 5–15; English transl., Automatic Documentation and Math. Linguistics **15** (1981), no. 3, 17–24.

2. Nuel D. Belnap, Jr., and Thomas B. Steel, Jr., *The logic of questions and answers*, Yale Univ. Press, New Haven, Conn., 1976.

3. M. Presburger, *Über die Vollständigkeit eines gewissen Systems der Arithmetik ganzer Zahlen, in welchem die Addition als einziger Operation hervortritt*, C. R. I Congr. Math. Pays Slaves (Warsaw, 1929), Książnica Atlas T.N.S.W., Warsaw, 1930, pp. 92–101, 395.

4. Elliott Mendelson, *Introduction to mathematical logic*, Van Nostrand, Princeton, N.J., 1964.

5. D. Hilbert and P. Bernays, *Grundlagen der Mathematik*. Vol. I, rev. ed., Springer-Verlag, 1968.

6. J. Richard Büchi, *Weak second-order arithmetic and finite automata*, Z. Math. Logik Grundlagen Math. **6** (1960), 66–92.

7. Yu. G. Penzin, *Arithmetical questions in number theory*, Algebraic Systems (A. I. Kokorin et al., editors), Irkutsk. Gos. Univ., Irkutsk, 1976, pp. 122–148. (Russian)

Translated by G. L. CHERLIN

Amer. Math. Soc. Transl.
(2) Vol. **140**, 1988

Application of Hausdorff Dimension in the Theory of Diophantine Approximations

V. I. BERNIK

§1. Introduction

Let S be some set of real numbers. For any $\delta > 0$ we define

$$m(\rho, \delta, S) = \inf \sum_{j=1}^{\infty} |I_j|^{\rho},$$

where the greatest lower bound is considered over all possible sequences of intervals $I_1, I_2, \dots$ of length $|I_j| \leq \delta$ that generate a covering of S. We set

$$m(\rho, S) = \sup_{\delta > 0} m(\rho, \delta, S).$$

Then

$$r = \inf\{\rho \colon m(\rho, S) = 0\}$$

is called the *Hausdorff dimension of the set* S, and we write $r = \dim S$. It is not hard to show that the dimension of a countable set equals zero and that $0 \leq \dim S \leq 1$ for any S.

We shall sometimes use the following property of dimension:

$$\dim \bigcup_{i=1}^{\infty} S_i = \sup_i \dim S_i.$$

In recent years this concept, which was introduced by Hausdorff [16], has often been applied in the analysis of various sets characterized by number-theoretic properties. We denote by $\mathscr{L}_n(w)$ the set of real number ω for which the inequality

$$|P(\omega)| < H^{-w} \tag{1}$$

1980 *Mathematics Subject Classification* (1985 *Revision*). Primary 11J69; Secondary 26C10, 30C15.

Translation of Acta Arith. **42** (1983), 219–253; MR **85k**: 11035.

has an infinite number of solutions in integral polynomials $P(x)$ of degree at most n. Here $H = \max_{0 \le i \le n} |a_i|$ is the height of the polynomial

$$P(x) = a_n x^n + \cdots + a_1 x + a_0. \tag{2}$$

At the end of the twenties Jarník [17] and Besicovitch [14] proved that $\dim \mathscr{L}_1(w) = 2/(w+1)$ for $w > 1$. Güting [15] strengthened this result.

For $n > 1$ the following is known. It follows from Minkowski's linear forms theorem that $\mathscr{L}_n(n) = R$. Sprindzhuk [6]–[9], in the solution of Mahler's problem, showed that for $w > n$ the set $\mathscr{L}_n(w)$ has Lebesgue measure zero. In 1970 A. Baker and Schmidt [11] proved the inequality

$$\frac{n+1}{w+1} \le \dim \mathscr{L}_n(w) < 2\frac{n+1}{w+1}. \tag{3}$$

The bulk of their paper was devoted to obtaining a lower bound for $\mathscr{L}_n(w)$. An upper bound was obtained by applying Wirsing's inequalities [20].

A number of considerations led to the conjecture [11] that for $w > n$

$$\dim \mathscr{L}_n(w) = (n+1)/(w+1). \tag{4}$$

For $n = 1$ this conjecture is valid by virtue of the results indicated above. For $n = 2$ Kasch and Volkmann [18] proved that $\dim \mathscr{L}_2(w) \le 3/(w+1)$. Together with inequality (3) this yields (4) for $\mathscr{L}_2(w)$. Finally, quite recently R. Baker [13] proved that $\dim \mathscr{L}_3(w) \le 4/(w+1)$ for $w > 3$ and that $\dim \mathscr{L}_n(w) \le (n+1)/(w+1)$ for $n \ge 4$ and $w > (n^2+n-3)/3$, which together with (3) yields (4) for $n = 3$ and $w > (n^2+n-3)/3$ for $n \ge 4$. For $(4n-3)/3 \le w \le (n^2+n-3)/3$, R. Baker obtained the estimate

$$\dim \mathscr{L}_n(w) \le \frac{n}{w+1-n/3}.$$

However, for $n < w < (4n-3)/3$ his method does not make it possible to obtain an estimate better than the trivial one ($\dim \mathscr{L}_n(w) \le 1$). The estimate $\dim \mathscr{L}_n(w) < 1$ can be obtained by applying Sprindzhuk's method of essential and unessential regions [6]–[9], [12]. Such results were obtained by the author and prepared for publication. However, shortly thereafter it became clear that the method allows further development, which led to a complete proof of (4).

THEOREM. *The Hausdorff dimension of the set* $\mathscr{L}_n(w)$ *for* $w > n$ *equals* $(n+1)/(w+1)$.

Obviously, if one takes inequality (3) into account, the theorem will follow from the inequality $\dim \mathscr{L}_n(w) \le (n+1)/(w+1)$. To establish this inequality is the goal of our subsequent investigations. The proof of the theorem is divided into five steps (§§4-8). In §2 the main concepts are introduced and auxiliary lemmas are proved. The next section is devoted to the proof of the main lemma—Lemma 12. In §4 some special cases of the theorem are examined (polynomials of the first and second degrees and cases of special distribution of the roots of polynomials). The next four sections are fundamental to the proof of the theorem. In §5 a proof of the theorem is given under the additional assumption that at least two polynomials from some set have no roots in common (Proposition 2). If this additional assumption does not hold, then we shall pass to a set of polynomials having a common factor. Although it is necessary to employ different arguments for different w, their goal is the same: to prove that a common factor has small degree and height and, at the same time, is sufficiently small at all points of some interval. We thus obtain a proof of the theorem for $n < w < 2n+1$ (§6, Proposition 3), for $w > 4n+3$ (§7, Proposition 4), and for $2n+1 \leq w \leq 4n+3$ (§8, Proposition 5).

A brief account of the proof of the theorem was given by the author in [2].

I express my deep thanks to Professor V. G. Sprindzhuk, who got me interested in the conjecture on the dimension of $\mathscr{L}_n(w)$ and rendered constant support in my efforts to prove it.

§2. Auxiliary lemmas

In this section we shall pass from the set of all integral polynomials of the form (2) to the set of polynomials irreducible in the field of rational numbers, having a definite coefficient and root structure.

We shall denote by $c(n)$ positive functions depending only on n and some positive number ε fixed in advance. We shall construct actions on $c(n)$ by the formal rules

$$c(n) + c(n) = c(n), \qquad c(n) \cdot c(n) = c(n),$$

which is taken to mean that the sum and product are again functions depending on n and ε.

If $P(x)$ is a polynomial of the form (2), then we shall denote by $H(P)$ the value of its coefficient largest in absolute value—the height of $P(x)$. If no ambiguity arises, instead of $H(P)$ we shall write simply H. It is clear that we can assume $H > H_0(n)$, where $H_0(n)$ is a sufficiently large natural number.

LEMMA 1. *Let $P_1(x), \ldots, P_l(x)$ be polynomials and $P(x) = P_1(x) \cdots P_l(x)$. Then*

$$c_1(n) H(P_1) \cdots H(P_l) < H(P) < c_2(n) H(P_1) \cdots H(P_l).$$

The proof can be found in [3] and [9].

LEMMA 2. *Let $w_1(\omega)$ and $v_1(\omega)$ be the least upper bounds of those w_1' and v_1' for which the inequalities*

$$|P(\omega)| < H(P)^{-w_1'}, \qquad |F(\omega)| < H(F)^{-v_1'}$$

have an infinite number of solutions in integral polynomials $P(x)$ and irreducible integral polynomials $F(x)$ respectively. Then $w_1(\omega) = v_1(\omega)$.

The proof is similar to the proof of the assertion in §5 of Chapter 1 in [9].

LEMMA 3. *Suppose that for some $w > n$ the inequality*

$$|P(\omega)| < H(P)$$

has an infinite number of solutions in irreducible integral polynomials $P(x)$ for some set $A(w)$, where $\dim A(w) \geq \delta$ $(0 \leq \delta \leq 1)$. Then the inequality

$$|Q(\omega)| < H(Q)^{-w}$$

has an infinite number of solutions for some set $B(w)$ in irreducible integral polynomials $Q(x)$ subject to the condition

$$\max_{0 \leq i \leq n} |a_i(Q)| \leq a_n(Q) = H(Q), \tag{5}$$

where the $a_i(Q)$ are the coefficients of $Q(x)$. Moreover, $\dim B(w) \geq \delta$.

The proof is similar to the proof of the assertion in §6 of Chapter 1 in [9]. An assertion similar to Lemma 3 was proved in Lemma 4 of [13].

We denote by $P_n(H)$ the class of irreducible polynomials $P(x)$ with condition (5) for which $a_n(P) = H$. Further, let $P_n = \bigcup_{H=1}^{\infty} P_n(H)$.

Let $P(x) \in P_n(H)$ and let $\kappa_1, \dots, \kappa_n$ be its roots. Then it is not hard to show that

$$|\kappa_i| \leq n, \qquad i = 1, \dots, n. \tag{6}$$

For any root κ_i of the polynomial $P(x)$ we construct an ordering of roots with respect to κ_i:

$$|\kappa_i - \kappa_{i2}| \leq |\kappa_i - \kappa_{i3}| \leq \dots \leq |\kappa_i - \kappa_{in}|. \tag{7}$$

It is clear that $\kappa_i, \kappa_{i2}, \dots, \kappa_{in}$ are some rearrangement of the roots $\kappa_1, \dots, \kappa_n$. For $i = 1$ we rewrite (7) in the form

$$|\kappa_1 - \kappa_2| \leq |\kappa_1 - \kappa_3| \leq \dots \leq |\kappa_1 - \kappa_n|, \tag{8}$$

dropping the index 1 from κ_{i1}.

LEMMA 4. *Let $A = \{A_1, \dots, A_n, \dots\}$ be a system of intervals A_i on the line and*

$$\sum_{i=1}^{\infty} |A_i|^\rho < \infty. \tag{9}$$

Let B be the set of real ω that belong to an infinite number of A_i. Then $\dim B \leq \rho$.

The proof is simple.

It follows from Lemma 4 that if we prove that $\mathscr{L}_n(w)$ can be covered by a system of intervals $S = \{S_1, \dots, S_n, \dots\}$ with the condition

$$\sum_{i=1}^{\infty} |S_i|^\rho < \infty$$

for $\rho = (n+1)/(w+1) + \varepsilon$ for any $\varepsilon > 0$, then the theorem will follow. We fix ε and set $T = [d(n)/\varepsilon]$ and $\varepsilon_1 = T^{-1}$, where $d(n)$ is a sufficiently large value that depends only on n.

We introduce the notation

$$|\kappa_i - \kappa_{ij}| = H^{-\mu_{ij}}, \qquad j = 2, \dots, n. \tag{10}$$

For $i = 1$ we write (10) in the form

$$|\kappa_1 - \kappa_j| = H^{-\mu_j}, \qquad j = 2, \dots, n. \tag{11}$$

We define the integers l_{ij} from the inequalities

$$l_{ij}/T \le \mu_{ij} < (l_{ij} + 1)/T, \qquad j = 2, \dots, n. \tag{12}$$

For $i = 1$ we write (12) in the form

$$l_j/T \le \mu_j < (l_j + 1)/T. \tag{13}$$

With each root κ_i of the polynomial $P(x)$ we associate the integral vector $\overline{s}_i = (l_{i2}, \dots, l_{in})$. We combine all polynomials $P(x) \in P_n(H)$ having the same vector $\overline{s}_i$ into one class $P_n(H, \overline{s}_i)$. For $i = 1$ we denote the class $P_n(H, \overline{s}_1)$ by $P_n(H, \overline{s})$.

LEMMA 5. *For any $i = 1, \dots, n$ the number of classes $P_n(H, \overline{s}_i)$ is finite and depends only on n and ε.*

Lemma 5 was proved in [1]. Since the number of classes $P_n(H, \overline{s})$ is finite, we shall assume that inequality (2) is being examined for some fixed vector $\overline{s}$.

Let us introduce one more bit of notation. Let $\mathscr{S}(\kappa_i)$ be the set of real numbers ω having the property

$$\min_{1 \le j \le n} |\omega - \kappa_j| = |\omega - \kappa_i|. \tag{14}$$

It is clear that all the $\mathscr{S}(\kappa_i)$ are intervals, the two outermost intervals having the form $(-\infty, \beta_1)$ and (β_2, ∞), where β_1 and β_2 are real algebraic numbers. We arrange the roots κ_i of the polynomials $P(x)$ in increasing order according to the value $\operatorname{Re} \kappa_i$. If κ_i is a complex root, then we write first the root for which $\operatorname{Im} \kappa_i > 0$. It is clear that if inequality (2) holds infinitely often for some $\omega \in R$, then ω belongs to an infinite number of $\mathscr{S}(\kappa_i)$ for some fixed i $(1 \le i \le n)$. We shall assume that $i = 1$, which, of course, does not lessen the generality.

We define the numbers p_i $(i = 1, \dots, n-1)$ as follows:

$$p_i = (l_{i+1} + \cdots + l_n)/T. \tag{15}$$

LEMMA 6. *Let $P(x) \in P_n(H)$ and $\omega \in \mathscr{S}(\kappa_1)$. Then*

$$|\omega - \kappa_1| \le 2^n \frac{|P(\omega)|}{|P'(\kappa_1)|}, \tag{16}$$

$$|\omega - \kappa_1| \le \min_{2 \le j \le n} \left(2^n \frac{|P(\omega)|}{|P'(\kappa_1)|} |\kappa_1 - \kappa_2| \cdots |\kappa_1 - \kappa_j| \right)^{1/j}. \tag{17}$$

Lemma 6 was proved, for example, in [1].

LEMMA 7. *Let $P(x) \in P_n(H, \overline{s})$. Then*

$$|P^{(l)}(\kappa_1)| < c(n) H^{1 - p_l}. \tag{18}$$

Lemma 7 was also proved in [1].

LEMMA 8. *Let $P(x)$ be a polynomial of degree n and height H. Suppose that the leading coefficient of $P(x)$ equals a_n and that $i_1, \dots, i_m$ are pairwise distinct natural numbers that assume the values $1, \dots, n$. Then for any m*

$$|\kappa_{i_1} \cdots \kappa_{i_m}| < c(n)\frac{H}{|a_n|},$$

where $\kappa_1, \dots, \kappa_n$ are the roots of $P(x)$.

Lemma 8 was proved in [10].

LEMMA 9. *Let $P(H)$ be the set of polynomials of degree at most l and height at most H, and let $|P(H)|$ be the number of elements in $P(H)$. If for any polynomial $P(x) \in P(H)$ the inequality $|P(\omega)| < H^{\mu}$ holds for all ω in some interval I, then*

$$|P(H)| < \begin{cases} c(l)H^{l+\mu}, & 0 \le \mu \le 1, \\ c(l)H^{l}, & \mu < 0. \end{cases}$$

PROOF. Suppose that $|P(\omega)| < H^{\mu}$ for some $P(x) \in P(H)$. Let b be some integer. Then polynomials of the form $P(\omega) + b$ can satisfy the inequality $|P(\omega) + b| < H^{\mu}$ only for $-2H^{\mu} < b < 2H^{\mu}$. There are at most $c(l)H^{1+\mu}$ polynomials $P(x)$ satisfying restrictions of this form for $0 \le \mu \le 1$, and there are $c(l)H^{l}$ of them for $\mu < 0$.

LEMMA 10. *Let $I \subset R$ be some interval and $B \subset I$ a measurable set of real numbers with $|B| \ge k^{-1}|I|$, where k is a natural number. Suppose that $|P(\omega)| < H^{-w}$ for all $\omega \in B$, where $w > 0$ and $\deg P(x) \le n$. Then for all $\omega \in I$*

$$|P(\omega)| < (3k)^n (n+1)^{n+1} H^{-w}.$$

PROOF. We divide the interval I into $3k(n+1)$ equal intervals $I_1, \dots, I_{3k(n+1)}$. Since $|B| \ge k^{-1}|I|$, in at least $3n+3$ intervals $\omega \in B$ occurs. We denote these intervals by $I_{s1}, \dots, I_{sl}$, $l \ge 3n+3$. We consider three consecutive intervals I_{sj}, $I_{s(j+1)}$, $I_{s(j+2)}$ and choose $\omega \in B \cap I_{s(j+1)}$. In this way we can choose $n+1$ points $\omega_1, \dots, \omega_{n+1}$ that satisfy the conditions

$$|\omega_i - \omega_j| \ge \frac{1}{3k(n+1)}|I| \quad (i \ne j)$$

and at which $|P(\omega_i)| < H^{-w}$. We represent $P(\omega)$ in the form of the Lagrange interpolation formula

$$P(\omega) = \sum_{l=1}^{n+1} c_l \frac{(\omega - \omega_1)\cdots(\omega - \omega_{l-1})(\omega - \omega_{l+1})\cdots(\omega - \omega_{n+1})}{(\omega_l - \omega_1)\cdots(\omega_l - \omega_{l-1})(\omega_l - \omega_{l+1})\cdots(\omega_l - \omega_{n+1})},$$

where c_l is the value of the polynomial $P(\omega)$ at the point $\omega = \omega_l$. Since $|\omega - \omega_l| \le I$ for all $\omega \in I$ and, by the condition of the lemma, $|c_l| < H^{-w}$, from the representation of $P(\omega)$ we obtain

$$|P(\omega)| < \sum_{l=1}^{n+1} H^{-w} \frac{|I|^n}{(3k(n+1))^{-n}|I|^n} = (3k)^n (n+1)^{n+1} H^{-w}.$$

LEMMA 11. *Let $P_n'(H,\bar{s})$ be the subset of $P_n(H,\bar{s})$ consisting of the $P(x)$ for which there exists an $\omega \in R$ that satisfies* (1). *Denote by $|P_n'(H,\bar{s})|$ the number of polynomials in the subset $P_n'(H,\bar{s})$, and suppose that for all polynomials $P(x) \in P_n(H,\bar{s})$ the value $p_1(P) = p_1$ satisfies the condition $0 \leq p_1 \leq 1$. Then*

$$|P_n'(H,\bar{s})| < c(n)H^{n-p_1+(n-1)\varepsilon_1}.$$

An assertion similar to Lemma 11 was proved in [13].

§3. Proof of the main lemma

LEMMA 12. *Let $\delta > 0$ be a real number, s a natural number, and $H = H(\delta, s)$ a sufficiently large real number. Further, let $P(x)$, $Q(x) \in Z[x]$ be two relatively prime polynomials with*

$$\max(H(P), H(Q)) = H^{\mu}, \qquad \deg P(x) \leq s, \quad \deg Q(x) \leq s.$$

If for all ω in some interval $I \subset (-s,s)$, $|I| = H^{-\eta}$, $\eta > 0$,

$$\max(|P(\omega)|, |Q(\omega)|) < H^{-\tau}, \qquad \tau > 0,$$

then

$$\tau + \mu + 2\max(\tau + \mu - \eta, 0) < 2\mu s + \delta.$$

PROOF. Let $\kappa_1(P), \dots, \kappa_{s_1}(P)$ be the roots of the polynomial $P(x)$, and let $\kappa_1(Q), \dots, \kappa_{s_2}(Q)$ be the roots of $Q(x)$. Here s_1 and s_2 are the degrees of $P(x)$ and $Q(x)$; $s_1, s_2 \leq s$. Let

$$\nu_i(P) = I \cap \mathscr{S}(\kappa_i(P)), \qquad i = 1, \dots, s_1,$$
$$\nu_j(Q) = I \cap \mathscr{S}(\kappa_j(Q)), \qquad j = 1, \dots, s_2.$$

It is possible that $\nu_i(P) = \varnothing$ and $\nu_j(Q) = \varnothing$ for some i and j. Since the number of roots of each polynomial does not exceed s, there exist i and j such that

$$|\nu_i(P)| \geq \frac{|I|}{s} = \frac{H^{-\eta}}{s}, \qquad |\nu_j(Q)| \geq \frac{|I|}{s} = \frac{H^{-\eta}}{s}.$$

We shall assume for simplicity of notation that $i = j = 1$. From inequality (17) for $j = n$ one can conclude that the distance from the roots $\kappa_1(P)$ and $\kappa_1(Q)$ to the interval I does not exceed $c(s)H^{-\tau/s}$. With respect to the root $\kappa_1(P)$ we construct an ordering of the remaining roots:

$$\begin{aligned} |\kappa_1(P) - \kappa_2(P)| &\leq |\kappa_1(P) - \kappa_3(P)| \leq \dots \leq |\kappa_1(P) - \kappa_t(P)| \\ &\leq s \leq |\kappa_1(P) - \kappa_{t+1}(P)| \leq \dots \leq |\kappa_1(P) - \kappa_{s_1}(P)|. \end{aligned} \tag{19}$$

Next, we denote $|\kappa_1(P) - \kappa_j(P)|$ by $H^{-m_j(P)}$, $j = 1, \dots, s_1$, and we set

$$p_i(P) = m_{i+1}(P) + \dots + m_t(P), \qquad i = 1, 2, \dots, t-1, \tag{20}$$
$$p_j'(P) = m_{t+j}(P) + \dots + m_{s_1}(P), \qquad j = 1, \dots, s_1 - t. \tag{21}$$

From the way t is defined it follows that for sufficiently large H all $m_k(P) \geq -\delta/s^2$, $k < t$. Let a_j be the coefficient largest in absolute value of the polynomial

$$P(x) = a_{s_1}x^{s_1} + \dots + a_1 x + a_0, \qquad 0 \leq j \leq s_1, \quad |a_{s_1}| = H^{\nu_1}, \quad 0 \leq \nu_1 \leq \mu.$$

Since

$$P(x) = a_{s_1}(x - \kappa_1) \cdots (x - \kappa_{s_1}),$$

there exists a collection of roots $\kappa_{i_1}, \dots, \kappa_{i_k}$ such that

$$|\kappa_{i_1} \cdots \kappa_{i_k}| > c(s_1) H^{\mu - \nu_1}. \tag{22}$$

Let us discard from this collection all roots less than $2s$ in absolute value. Inequality (22) is then preserved with a different value of $c(s_1)$. Without loss of generality we shall assume that all roots $\kappa_{i_1}, \dots, \kappa_{i_k}$ have absolute values that exceed $2s$. From the conditions of the theorem and (16) we conclude that

$$|\omega - \kappa_1(P)| < c(s_1) H^{-\tau - \nu_1 + p_1(P) + p_t'(P)}.$$

It is clear that none of the roots $\kappa_2(P), \dots, \kappa_t(P)$ occurs in the set $\{\kappa_{i_1}, \dots, \kappa_{i_k}\}$. On the other hand, since $|\kappa_1(P)| \leq s$,

$$|\kappa_{i_j}(P) - \kappa_1(P)| \geq \frac{1}{2} |\kappa_{i_j}(P)|, \qquad j = 1, \dots, k.$$

Therefore,

$$|\kappa_1(P) - \kappa_{t+1}(P)| \cdots |\kappa_1(P) - \kappa_{s_1}(P)| = H^{p_t'(P)} > c(s_1) H^{\mu - \nu_1}$$

and

$$|\omega - \kappa_1(P)| < c(s_1) H^{-\tau - \mu + p_1(P)}. \tag{23}$$

By analogy with (19) we order all roots of the polynomial

$$Q(x) = b_{s_2} x^{s_2} + \cdots + b_1 x + b_0, \qquad |b_{s_2}| = H^{\nu_2}, \quad 0 \leq \nu_2 \leq \mu,$$

with respect to the root $\kappa_1(Q)$, and then we define $p_i(Q)$ and $p_j'(Q)$, $i = 1, \dots, t_2 - 1$ and $j = 1, \dots, s_2 - t_2$, as in (20) and (21). After doing this it is not hard to obtain the following inequality, which is analogous to (23):

$$|\omega - \kappa_1(Q)| < c(s_2) H^{-\tau - \mu + p_1(Q)}. \tag{24}$$

Reasoning as in the derivation of (23), from (17) for $\omega \in \nu_1(P)$ we obtain

$$|\omega - \kappa_1(P)| < c(s) \min_{1 \leq j \leq t-1} H^{-(\tau + \mu - p_j(P))/j}. \tag{25}$$

Suppose that the minimum on the right side of (25) is attained for $j = j'$. We now take $\omega \in \nu_1(Q)$. Then again with the help of (17) we obtain

$$|\omega - \kappa_1(Q)| < c(s) \min_{1 \leq k \leq t_2 - 1} H^{-(\tau + \mu - p_k(Q))/k}. \tag{26}$$

Suppose that the minimum on the right side of (26) is attained for $k = j''$. From the definition of j' we obtain, for any i, $1 \leq i \leq s_1$,

$$H^{-(\tau + \mu - p_{j'}(P))/j'} \leq H^{-(\tau + \mu - p_i(P))/i},$$

which implies the inequality

$$i(\tau + \mu - p_{j'}(P)) \geq j'(\tau + \mu - p_i(P)). \tag{27}$$

Let $i < j'$. It then follows from (27) that

$$m_{i+1}(P) \geq \frac{\tau + \mu - p_{j'}(P)}{j'}. \tag{28}$$

For $i > j'$, from (27) we obtain

$$m_i(P) \leq \frac{\tau + \mu - p_i(P)}{i}, \qquad i = j' + 1, \ldots, s_1. \tag{29}$$

From the definition of j'' one can obtain formulas for the polynomial $Q(\omega)$ analogous to (28) and (29):

$$m_{t+1}(Q) \geq \frac{\tau + \mu - p_{j''}(Q)}{j''}, \qquad t = 1, \ldots, j'' - 1, \tag{30}$$

$$m_t(Q) \leq \frac{\tau + \mu - p_t(Q)}{t}, \qquad t = j'' + 1, \ldots, s_2. \tag{31}$$

From the definition of $\nu_1(P)$ we obtain $|\nu_1(P)| > c(s)H^{-\eta}$. Since the value $|\omega - \kappa_1(P)|$ for a suitable choice of $\omega \in \nu_1(P)$ can be made at least $\frac{1}{2}|\nu_1(P)|$, from (23) we have

$$H^{-(\tau+\mu-p_j(P))/j} \geq H^{-\eta}, \qquad \eta \geq \frac{\tau + \mu - p_j(P)}{j}. \tag{32}$$

An inequality analogous to (32) can also be obtained for $Q(\omega)$:

$$\eta \geq \frac{\tau + \mu - p_t(Q)}{t}, \qquad t = 1, \ldots, s_2. \tag{33}$$

From the definition of j' for any $i = 1, \ldots, s_1$ we have

$$H^{-(\tau+\mu-p_{j'}(P))/j'} \leq H^{-(\tau+\mu-p_i(P))/i},$$

whence and from (29) for $i > j'$ we get

$$m_i(P) \leq \frac{\tau + \mu - p_{j'}(P)}{j'}, \qquad i = j' + 1, \ldots, s_1. \tag{34}$$

An inequality analogous to (34) is also valid for $Q(\omega)$:

$$m_t(Q) \leq \frac{\tau + \mu - p_{j''}(Q)}{j''}, \qquad t = j'' + 1, \ldots, s_2. \tag{35}$$

We assume for definiteness that

$$\frac{\tau + \mu - p_{j''}(Q)}{j''} \geq \frac{\tau + \mu - p_{j'}(P)}{j'}. \tag{36}$$

We now proceed to estimate the differences between the roots of the polynomials $P(\omega)$ and $Q(\omega)$. We first estimate the difference

$$|\kappa_1(P) - \kappa_1(Q)| \leq |\kappa_1(P) - \omega'| + |\omega' - \omega''| + |\omega'' - \kappa_1(Q)|. \tag{37}$$

In (37) we have $\omega' \in \nu_1(P) \subset I$ and $\omega'' \in \nu_1(Q) \subset I$, and so $|\omega' - \omega''| \leq H^{-\eta}$. The differences $|\omega' - \kappa_1(P)|$ and $|\omega'' - \kappa_1(Q)|$ are estimated in (23) and (24). Hence,

$$|\kappa_1(P) - \kappa_1(Q)| < c(s)(H^{-(\tau+\mu-p_{j'}(P))/j'} + H^{-(\tau+\mu-p_{j''}(Q))/j''} + H^{-\eta}). \tag{38}$$

Inequalities (32), (33), and (36) make it possible to rewrite (38) in the form

$$|\kappa_1(P) - \kappa_1(Q)| < c(s)H^{-(\tau+\mu-p_{j'}(P))/j'}. \tag{39}$$

Furthermore, from the inequality

$$\eta \geq \frac{\tau + \mu - p_{j''}(Q)}{j''},$$

which is obtained from (33) for $t = j''$ and (30) and (36), it follows that

$$\begin{aligned}|\kappa_i(Q) - \kappa_1(P)| &\leq |\kappa_i(Q) - \kappa_1(Q)| + |\kappa_1(Q) - \kappa_1(P)| \\ &\leq c(s)H^{-(\tau+\mu-p_{j'}(P))/j'}\end{aligned} \tag{40}$$

for $i = 1, \dots, j''$. Now from (40) and (28) for any $1 \leq i \leq j''$ and $1 \leq j \leq j'$ we obtain

$$\begin{aligned}|\kappa_i(Q) - \kappa_j(P)| &\leq |\kappa_i(Q) - \kappa_1(P)| + |\kappa_1(P) - \kappa_j(P)| \\ &< c(s)H^{-(\tau+\mu-p_{j'}(P))/j'}.\end{aligned} \tag{41}$$

From (41) we conclude that

$$\prod_{1\leq i\leq j''} \prod_{1\leq j\leq j'} |\kappa_i(Q) - \kappa_j(P)| < c(s)H^{-j''(\tau+\mu-p_{j'}(P))}. \tag{42}$$

On the other hand, by (34) for any $i = 1, \dots, j''$

$$\prod_{j>j'} |\kappa_i(Q) - \kappa_j(P)| < c(s)H^{-p_{j'}(P)},$$

whence

$$\prod_{1\leq i\leq j''} \prod_{j>j'} |\kappa_i(Q) - \kappa_j(P)| < c(s)H^{-j''p_j(P)}. \tag{43}$$

It follows from (42) and (43) that

$$\prod_{1\leq j\leq s_1} \prod_{1\leq i\leq j''} |\kappa_i(Q) - \kappa_j(P)| < c(s)H^{-j''(\tau+\mu)}. \tag{44}$$

Since $P(\omega)$ and $Q(\omega)$ have no roots in common, $|R(P,Q)| \geq 1$, where R is the resultant of $P(\omega)$ and $Q(\omega)$. Using (44), we obtain

$$\begin{aligned}1 \leq |R(P,Q)| &\leq H^{(\nu_1+\nu_2)s} \prod_{1\leq i\leq s_2} \prod_{1\leq j\leq s_1} |\kappa_i(Q) - \kappa_j(P)| \\ &= H^{(\nu_1+\nu_2)s} \prod_{1\leq i\leq t_2} \prod_{1\leq j\leq t} |\kappa_i(Q) - \kappa_j(P)| \prod_{\substack{i>t_2 \\ 1\leq j\leq s_1}} |\kappa_i(Q) - \kappa_j(P)| \\ &\quad \times \prod_{1\leq i\leq s_2} \prod_{j>t} |\kappa_i(Q) - \kappa_j(P)|.\end{aligned} \tag{45}$$

By Lemma 8 the last two products can be bounded from above by the value $c(s)H^{s_1(\mu-\nu_1)+s_2(\mu-\nu_2)}$, and so (45) can be rewritten in the form

$$1 \leq c(s)H^{2\mu s - j''(\tau+\mu)}.$$

The inequality $j''(\tau+\mu) \leq 2\mu s + \delta$ for $j \geq 3$, which is obtained from this, is stronger than the one we need to prove. Let us consider the case $j''=2$. If

$$m_3(Q) \geq \frac{\tau+\mu-p_{j'}(P)}{j'}, \tag{46}$$

then for $1 \leq j \leq j'$

$$|\kappa_j(P)-\kappa_3(Q)| < |\kappa_j(P)-\kappa_1(Q)| + |\kappa_1(Q)-\kappa_3(Q)|$$

and from (42) and (46) we obtain

$$\prod_{1\leq j\leq j'} |\kappa_j(P)-\kappa_3(Q)| < c(s)H^{-(\tau+\mu-p_{j'}(P))}. \tag{47}$$

Using (34) and (46), for $j > j'$ we obtain

$$\begin{aligned}|\kappa_j(P)-\kappa_3(Q)| &< |\kappa_j(P)-\kappa_1(P)| + |\kappa_1(P)-\kappa_1(Q)| + |\kappa_1(Q)-\kappa_3(Q)| \\ &< c(s)H^{-m_j(P)},\end{aligned}$$

and so

$$\prod_{j>j'} |\kappa_j(P)-\kappa_3(Q)| < c(s)H^{p_{j'}(P)}. \tag{48}$$

It now follows from (47) and (48) that

$$\prod_{1\leq j\leq s_1} |\kappa_j(P)-\kappa_3(Q)| < c(s)H^{-(\tau+\mu)}. \tag{49}$$

From (44) and (45) we obtain, for $j''=2$,

$$\prod_{1\leq i\leq s_2}\prod_{1\leq j\leq s_1} |\kappa_i(Q)-\kappa_j(P)| < c(s)H^{-(\tau+\mu)} \tag{50}$$

which makes it possible to obtain the inequality $3(\tau+\mu) \leq 2\mu s$, which is enough to prove the lemma.

If, contrary to (46),

$$m_3(Q) < \frac{\tau+\mu-p_{j'}(P)}{j'}, \tag{51}$$

then from (51) and the inequalities

$$|\kappa_1(P)-\kappa_i(Q)| \leq |\kappa_1(P)-\kappa_1(Q)| + |\kappa_1(Q)-\kappa_i(Q)|$$

for any $i \geq 3$ we obtain

$$\prod_{3\leq i\leq s_2} |\kappa_1(P)-\kappa_i(Q)| < c(s)H^{-p_2(Q)}. \tag{52}$$

Inequality (52), together with (44), leads to

$$\prod_{1\leq i\leq s_2}\prod_{1\leq j\leq s_1} |\kappa_i(Q)-\kappa_j(P)| < c(s)H^{-2(\tau+\mu)-p_2(Q)}. \tag{53}$$

We use (33) for $t=2$. We obtain $p_2(Q) \geq \tau+\mu-2\eta$. Therefore, the exponent in (53) can be replaced by $-(\tau+\mu)-2(\tau+\mu-\eta)$, which also leads to a proof of the lemma. Moreover, it is clear that $p_2(Q)$ can be replaced by $\tau+\mu-2\eta$ only when $\tau+\mu-2\eta\geq 0$. The case $j''=2$ has been examined completely.

For $j''=1$ we reason as follows. Let

$$m_2(Q) > \frac{\tau+\mu-p_{j'}(P)}{j'}, \qquad m_3(Q) > \frac{\tau+\mu-p_{j'}(P)}{j'}. \tag{54}$$

Then, by analogy with the case of (49), we obtain

$$\prod_{1\leq i\leq s_1}\prod_{2\leq j\leq 3}|\kappa_i(P)-\kappa_j(P)| < c(s)H^{-2(\tau+\mu)}$$

which, together with (44), yields the following inequality, which is analogous to (50):

$$\prod_{1\leq i\leq s_1}\prod_{1\leq j\leq s_2}|\kappa_i(P)-\kappa_j(Q)| < c(s)H^{-3(\tau+\mu)}.$$

If now

$$m_2(Q) > \frac{\tau+\mu-p_{j'}(P)}{j'}, \qquad m_3(Q) \leq \frac{\tau+\mu-p_{j'}(P)}{j'}, \tag{55}$$

then the first inequality in (55) leads to

$$\prod_{1\leq i\leq s_1}|\kappa_i(P)-\kappa_2(Q)| < c(s)H^{-(\tau+\mu)}$$

and the second leads to (52). The fulfillment of both inequalities, as we have already seen, leads to a proof of the lemma. It remains to analyze the case $m_2(Q)\leq(\tau+\mu-p_{j'}(P))/j'$, which can be treated in the same way as (51). We obtain the following inequality, which is analogous to (52):

$$\prod_{2\leq j\leq s_2}|\kappa_1(P)-\kappa_j(Q)| < c(s)H^{-p_1(Q)}. \tag{56}$$

If, furthermore, $m_2(P) < m_2(Q)$, then for $j\geq 2$ we obtain

$$\begin{aligned}|\kappa_j(P)-\kappa_2(Q)| &\leq |\kappa_j(P)-\kappa_1(P)|+|\kappa_1(P)-\kappa_1(Q)|+|\kappa_1(Q)-\kappa_2(Q)|\\ &< c(s)H^{-m_j(P)},\end{aligned}$$

whence

$$\prod_{2\leq j\leq s_1}|\kappa_j(P)-\kappa_2(Q)| < c(s)H^{-p_1(P)}. \tag{57}$$

But if $m_2(P)\geq m_2(Q)$, then for $i\geq 2$ we have

$$\begin{aligned}|\kappa_2(P)-\kappa_i(Q)| &\leq |\kappa_2(P)-\kappa_1(P)|+|\kappa_1(P)-\kappa_1(Q)|+|\kappa_1(Q)-\kappa_i(Q)|\\ &< c(s)H^{-m_i(Q)},\end{aligned}$$

whence

$$\prod_{2\leq i\leq s_2}|\kappa_2(P)-\kappa_i(Q)| < c(s)H^{-p_1(Q)}. \tag{58}$$

Inequalities (40), (52), (53), and (54) lead to

$$\prod_{1\le i\le s_2}\prod_{1\le j\le s_1}|\kappa_j(P)-\kappa_i(Q)|<c(s)H^{-\tau-\mu-p_1(Q)-\min(p_1(Q),p_1(P))}. \tag{59}$$

From the inequalities $\eta\ge\tau+\mu-p_1(Q)$ and $\eta\ge\tau+\mu-p_1(P)$ we obtain

$$\min(p_1(P),p_1(Q))\ge\tau+\mu-\eta. \tag{60}$$

If $\eta>\tau+\mu$, then (60) can be replaced by the trivial inequality

$$\min(p_1(P),p_1(Q))\ge-\delta/s. \tag{61}$$

Bearing in mind (60) and (61), one can rewrite inequality (59) in the form

$$\prod_{1\le i\le s_2}\prod_{1\le j\le s_1}|\kappa_j(P)-\kappa_i(Q)|<c(s)H^{-\tau-\mu-2\max(\tau+\mu-\eta,0)},$$

which, obviously, proves the lemma.

We note that if η is considerably less than $\tau+\mu$, then Lemma 12 can be strengthened by adding to the left side of the inequality $\tau+\mu+2\max(\tau+\mu-\eta,0)\le2\mu s+\delta$ terms of the form $2\max(\tau+\mu-k\eta,0)$, $k=2,\dots,\min(s_1,s_2)$.

The following lemma can be proved by analogy with Lemma 12.

LEMMA 13. *Let $\delta>0$ be a real number, s a natural number, and $H=H(\delta,s)$ a sufficiently large real number. Suppose that two relatively prime polynomials $P(x)$, $Q(x)\in Z[x]$ have degrees $s_1\le s$ and $s_2\le s$ and heights H^{λ_1} and H^{λ_2} respectively. Suppose that for all ω in some interval $I\subset(-s,s)$, $|I|=H^{-\eta}$, $\eta>0$, the following inequalities hold:*

$$|P(\omega)|<H^{-\tau_1},\qquad|Q(\omega)|<H^{-\tau_2},\qquad\tau_1>0,\quad\tau_2>0.$$

If $\tau_1+\lambda_1\ge2(\tau_2+\lambda_2)$, then

$$\lambda_1s_2+\lambda_2s_1>\tau_2+\lambda_2+\min(\tau_2+\lambda_2,\tau_1+\lambda_1-\eta)+\delta.$$

LEMMA 14. *Suppose that $P(x)\in Z[x]$ has degree at most l and height at most H^λ. Suppose that for all ω in some interval I the inequality $|P(\omega)|<H^{-v}$ holds, where $v>3\lambda l$. Then there exists a divisor $d(x)$ of the polynomial $P(x)$ ($\deg d(x)=l_1$ and $H(d(x))=H^{\lambda_1}$) which is a power of a polynomial, irreducible in the rational field, that satisfies, for all $\omega\in I$, the conditions*

$$|d(x)|<c(l)H^{-v+\lambda l},\qquad l_1\le l,\qquad H^{\lambda_1}<c(l)H^{\lambda}. \tag{62}$$

Lemma 14, both in its formulation and in its method of proof, is similar to a lemma of Gel'fond ([3], Chapter III, §4, Lemma 6).

Let $P(x)=d_1(x)d_2(x)$ and let $|d_1(\omega)|>|d_2(\omega)|$ for some set $B_1\subset I$, $|B_1|\ge\frac12|I|$. Let us show that, for any subset $B\subset B_1$ with measure at least $\frac12|B_1|$,

$$|d_1(\omega)|>H^{-\lambda l}. \tag{63}$$

We assume that there exists a set $B_2\subset B_1$, $|B_2|\ge\frac12|B_1|$, for all points of which the inequality does not hold. Then in the set B_2, $|B_2|\ge\frac14|I|$, we have the inequality opposite to (63) and the inequality $|d_2(\omega)|<H^{-\lambda l}$. We

denote the degrees and heights of the polynomials $d_1(\omega)$ and $d_2(\omega)$ by l_1 and l_2 and by H^{λ_1} and H^{λ_2} respectively. Then, since $l_1 + l_2 \le l$ and, by Lemma 1, $H^{\lambda_1+\lambda_2} < c(l)H^{\lambda}$, we have

$$H^{\lambda_1 l_2+\lambda_2 l_1} \le c(l)H^{\lambda l}.$$

Using Lemma 12, we obtain a contradiction. This means that inequality (63) holds in any set $B_3 \subset B_1$, $|B_3| \ge \frac{1}{4}|I|$, and hence

$$|d_2(\omega)| < H^{-v+\lambda l}$$

for all $\omega \in B_3$. Then by Lemma 10 the following inequality holds for all $\omega \in I$:

$$|d_2(\omega)| < c(l)H^{-v+\lambda l}. \tag{64}$$

We now represent $P(x)$ as a product of powers of distinct irreducible polynomials:

$$P(x) = d_1(x)\cdots d_s(x).$$

It is clear that there exists a constant $c(l) > 0$ such that in a set B_4, $|B_4| > c(l)|I|$, the following inequalities hold:

$$|d_{i_1}(\omega)| \le |d_{i_2}(\omega)| \le \cdots \le |d_{i_s}(\omega)|. \tag{65}$$

Since the inequality

$$|d_{i_1}(\omega)\cdots d_{i_k}(\omega)| > |d_{i_{k+1}}(\omega)\cdots d_{i_s}(\omega)|$$

cannot hold for $k \ge s/2$ by virtue of (65), there exists a $\nu < s/2$ such that

$$\begin{aligned} |d_{i_1}(\omega)\cdots d_{i_{\nu-1}}(\omega)| &> |d_{i_\nu}(\omega)\cdots d_{i_s}(\omega)|, \\ |d_{i_1}(\omega)\cdots d_{i_\nu}(\omega)| &< |d_{i_{\nu+1}}(\omega)\cdots d_{i_s}(\omega)|. \end{aligned} \tag{66}$$

But then for $\omega \in I$ as in (64) we have

$$|d_{i_\nu}(\omega)\cdots d_{i_s}(\omega)| < c(l)H^{-v+\lambda l},$$

and, on the other hand, as in (63), for $\omega \in B_4$

$$|d_{i_{\nu+1}}(\omega)\cdots d_{i_s}(\omega)| > H^{-\lambda l}.$$

Therefore, for $\omega \in B_4$ from (66) we obtain

$$|d_{i_1}(\omega)| \le |d_{i_\nu}(\omega)| < c(l)H^{-v+2\lambda l}, \tag{67}$$

but since $|B_4| > c(l)|I|$, by Lemma 10 an inequality analogous to (67) is also valid for all $\omega \in I$.

The polynomial $t(x) = P(x)d_{i_1}^{-1}(x)$ is relatively prime to $d_{i_1}(x)$. If we assume that for some set B_5, $|B_5| \ge \frac{1}{2}|I|$,

$$|t(\omega)| < H^{-\lambda l}, \tag{68}$$

then we obtain a contradiction, since by $v > 3\lambda l$ we have

$$|d_{i_1}(\omega)| < c(l)H^{-\lambda l}.$$

Therefore, the inequality opposite to (68) holds in any set B_6 with measure greater than $\frac{1}{2}|I|$. This leads to the inequality

$$|d_{i_1}(\omega)| < c(l)H^{-v+\lambda l},$$

which is valid for all $\omega \in B_5$ and, by Lemma 10, to the inequality

$$|d_{i_1}(\omega)| < c(l)H^{-v+\lambda l},$$

which is valid for all $\omega \in I$.

§4. Special cases of the theorem

We note that the theorem to be proved has already been proved for polynomials of first [14], [19], second [18], and third [13] degrees. With the use of Lemma 11 it is not hard to prove the theorem in the case when $0 \le p_1 \le 1$. Furthermore, if the vector $\bar{s}$ is such that

$$l_2/T > (n - p_2)/2, \tag{69}$$

then the proof of the theorem is simple and was carried out in [13]. We shall state all of the proved special cases of the theorem as a separate proposition.

PROPOSITION 1. $\dim \mathscr{L}_n(w) \le (n+1)/(w+1)$ *for* $n \le 3$. *For any* $n > 3$ *if the vector* $\bar{s}$ *satisfies the condition* $0 \le p_1 \le 1$ *or* $l_2/T > (n-p_2)/2$, *then also,* $\dim \mathscr{L}_n(w) \le (n+1)/(w+1)$.

Bearing Proposition 1 in mind, we assume in what follows that

$$n > 3, \qquad p_1 > 1, \qquad l_2/T \le (n - p_2)/2.$$

We make an inductive assumption: we assume that the theorem has been proved for all polynomials of degree less than n. This means that if we find that the inequality $|P(\omega)| < H^{-v}$ holds infinitely often for some set $B \subset R$ in polynomials $P(\omega)$ of degree $l \le n-1$, then we shall conclude that $\dim B \le (l+1)/(v+1)$.

§5. The nondegenerate case

In what follows we shall assume that $\omega \in (-n-1, n+1)$, since otherwise with the help of (6), (16), and (17) we can prove that inequality (2) cannot hold for any $w > 0$. Suppose, contrary to (69), that

$$\frac{l_2}{T} \le \frac{n - p_2}{2}. \tag{70}$$

We divide the interval $(-n-1, n+1)$ into equal parts by intervals of length $H^{-n+p_1+k-1-2n\varepsilon_1}$, where $k \ge 1$ is a real number that will be chosen a bit later. We shall say that $P(\omega)$ belongs to I if $\exists\ \omega \in I$, $|P(\omega)| < H^{-w}$. We denote by $P(I)$ the set of polynomials that belong to the same interval I. If a polynomial belongs to two intervals I (it is easy to show that three is impossible), then either one is chosen for this correspondence. Let $|P(I)|$ be the number of polynomials that belong to I, and let $|P(I)| < c(n)H^{k+\theta-1}$ for any I, where

$$\theta = p_1\left(1 - \frac{n+1}{w+1}\right). \tag{71}$$

Then the total number of polynomials $P(\omega) \in P_n(H, \overline{s})$ does not exceed $c(n)H^{n-p_1+\theta+2n\varepsilon_1}$, since the number of intervals I does not exceed $c(n)H^{n-p_1+k+1+2n\varepsilon_1}$. From (18) we find that all $\omega \in \mathscr{S}(\kappa_1)$ for which (2) holds are located inside an interval whose length does not exceed $c(n)H^{-w-1+p_1+n\varepsilon_1}$.

Since the series

$$\sum_{H=1}^{\infty} H^{n-p_1+\theta+2n\varepsilon_1}(H^{-w-1+p_1+n\varepsilon_1})^{(n+1)/(w+1)+\varepsilon}$$

$$< c(n)\sum_{H=1}^{\infty} H^{-1+\theta-p_1(1-(n+1)/(w+1))-\varepsilon/2}$$

converges, $\dim \mathscr{L}_n(w) \leq (n+1)/(w+1)$ and the theorem is proved. Hence, either we obtain a proof of the theorem or it remains to assume that there exist intervals I to which more than $c(n)H^{k+\theta-1}$ polynomials belong. Let us consider one such interval I. We represent k in the form $k = [k] + \{k\}$. We fix an integer i, $0 \leq i \leq [k]-1$. We shall associate to the same class two polynomials in the set $P_n(H, \overline{s})$,

$$P_1(x) = Hx^n + a_{n-1}^{(1)}x^{n-1} + \cdots + a_1^{(1)}x + a_0^{(1)},$$
$$P_2(x) = Hx^n + a_{n-1}^{(2)}x^{n-1} + \cdots + a_1^{(2)}x + a_0^{(2)},$$

that belong to the interval I if

$$\begin{gathered} a_{n-1}^{(1)} = a_{n-1}^{(2)}, \dots, a_{n-[k]+i+1}^{(1)} = a_{n-[k]+i+1}^{(2)}, \\ b_j H^{1-i/(n-[k]+i)} \leq a_j^{(1)}, \qquad a_j^{(2)} < (b_j+1)H^{1-i/(n-[k]+i)}, \\ b_j \in Z, \qquad -H^{i/(n-[k]+i)} \leq b_j < H^{i/(n-[k]+i)} - 1, \qquad j = 1, \dots, n-[k]+i. \end{gathered} \tag{72}$$

Let us apply Dirichlet's pigeonhole principle. Since the number of distinct classes does not exceed $c(n)H^{[k]-1}$, among $c(n)H^{k+\theta-1}$ polynomials there exist at least $c(n)H^{\theta+\{k\}}$ that belong to the same class. We number these polynomials $P_0(x), P_1(x), \dots, P_m(x)$, $m = c(n)H^{\theta}$. We form the new polynomials

$$t_1(x) = P_1(x) - P_0(x), \dots, t_m(x) = P_m(x) - P_0(x).$$

It is clear that all m polynomials have degree at most $n-[k]+i$ and that all of their coefficients, except possibly for the free term, do not exceed $2H^{1-i/(n-[k]+i)}$. However, since the $t_i(x)$, $1 \leq i \leq m$, assume small values in I, the free coefficient of the polynomials $t_i(x)$ cannot exceed $c(n)H^{1-i/(n-[k]+i)}$. Thus, we have

$$\deg t_j(x) \leq n-[k]+i, \qquad H(t_j(x)) < c(n)H^{1-i/(n-[k]+i)}. \tag{73}$$

All of the polynomials $t_j(x)$ are distinct, since otherwise the polynomials $P_j(x)$ would coincide. Let $P(x) \in P(I)$. We expand $P(\omega)$ in a neighborhood of the root κ_1 into a Taylor series

$$P(\omega) = P'(\kappa_1)(\omega-\kappa_1) + \frac{1}{2}P''(\kappa_1)(\omega-\kappa_1)^2 + \cdots + \frac{1}{n!}P^{(n)}(\kappa_1)(\omega-\kappa_1)^n. \tag{74}$$

Since $P(\omega)$ belongs to the interval I, there exists a point $\omega(P) \in I$ such that $|P(\omega(P))| \le H^{-w}$. From (16) we obtain

$$|\omega(P) - \kappa_1| \le 2^n H^{-w-1+p_1+n\varepsilon_1}.$$

If ω_1 is any other point of I, then, bearing in mind that $w > n$, we obtain

$$|\omega_1 - \kappa_1| \le |\omega_1 - \omega(P)| + |\omega(P) - \kappa_1| < c(n)H^{-n+p_1+k-1-2n\varepsilon_1}. \tag{75}$$

From (75) and Lemma 7 we obtain

$$|P'(\kappa_1)||\omega_1 - \kappa_1| < c(n)H^{-(n-k)-2n\varepsilon_1}.$$

If $l_2/T + p_1 + k \le n + 1 + 2n\varepsilon_1$, then

$$|P''(\kappa_1)||\omega_1 - \kappa_1|^2 < c(n)H^{-(n-k)-2n\varepsilon_1}.$$

For any $i \ge 3$

$$|P^{(i)}(\kappa_1)||\omega_1 - \kappa_1|^i < c(n)H^{1-p_i-i(n-p_1-k+1)-2ni\varepsilon_1} \le c(n)H^{-(n-k)-2n\varepsilon_1}.$$

Therefore, for all $\omega \in I$,

$$|P(\omega)| < c(n)H^{-(n-k)-2n\varepsilon_1}$$

and hence

$$|t_i(\omega)| < c(n)H^{-(n-k)-2n\varepsilon_1}. \tag{76}$$

We set

$$k = n + 1 - l_2/T - p_1 + (n+1)\varepsilon_1. \tag{77}$$

It follows from (70) that $k \ge 1$. In (72) and (73) we take $i = 0$. If among the polynomials $t_1(\omega), \dots, t_m(\omega)$ there are at least two without roots in common, then we apply Lemma 12. Here

$$\tau = n - k + 2n\varepsilon_1 = l_2/T + p_1 - 1 + (n-1)\varepsilon_1, \quad \mu = 1,$$
$$s = n - [k] = l_2/T + p_1 + \{k\} - 1 - (n-1)\varepsilon_1,$$
$$\eta = n - k - p_1 + 1 = l_2/T - (n-1)\varepsilon_1.$$

We obtain a contradiction. Thus, we have proved the following proposition.

PROPOSITION 2. *If among the polynomials $t_1(x), \dots, t_m(x)$ there are at least two without roots in common, then* $\dim \mathscr{L}_n(w) \le (n+1)/(w+1)$.

To complete the proof it remains to examine the case when among the polynomials $t_1(\omega), \dots, t_m(\omega)$ there are not at least two without roots in common.

§6. The degenerate case, $n < w < 2n + 1$

LEMMA 15. *Suppose that $N = c_1(n)H^\tau$ ($\tau > 0$) polynomials $t_j(\omega)$ belong to a given interval I. Then among them one can always choose $M = c_2(n)H^\tau$ polynomials $t_j(\omega)$, $j = 1, \dots, M$, that are representable in the form*

$$t_j(\omega) = k_j(\omega)d(\omega), \tag{78}$$

where among the polynomials $k_j(\omega)$ there are at least two without roots in common.

PROOF. We fix some polynomial

$$t_0(\omega) = P_1^{a_1}(\omega)\cdots P_s^{a_s}(\omega),$$

where the $P_i(\omega)$, $i = 1,\dots,s$, are irreducible polynomials. If among the N polynomials $t_i(\omega)$ there exist $\geq N/2$ having no roots in common with $t_0(\omega)$, then we obtain (78) with $d = 1$. If not, then among $c(n)H^\tau$ polynomials $t(\omega)$ there exist $M_1 = c(n)H^\tau$ polynomials having a factor $P_i(\omega)$, $i = 1,\dots,s$, which we denote by $d_1(\omega)$, in common with $t_0(\omega)$. Then

$$t_i(\omega) = k_i(\omega)d_1(\omega), \qquad i = 0,\dots,M_1. \tag{79}$$

If among the M_1 polynomials $k_i(\omega)$ there exist $\geq M_1/2$ having no roots in common with $t_0(\omega)d_1^{-1}(\omega)$, we arrive at (78) with $d(\omega) = d_1(\omega)$. If not, then there exist $M_2 = c(n)H^\tau$ polynomials having some factor $P_i(\omega)$, $i = 1,\dots,s$, which we denote by $d_2(\omega)$, in common with $t_0(\omega)d_1^{-1}(\omega)$. We arrive at the equalities

$$t_i(\omega) = k_i(\omega)d_1(\omega)d_2(\omega).$$

If it now turns out that among the polynomials $k_i(\omega)$, $i = 1,\dots,M_2$, there exist at least $M_2/2$ having no roots in common with $t_0(\omega)d_1^{-1}(\omega)d_2^{-1}(\omega)$, then we have obtained (78) with $d(\omega) = d_1(\omega)d_2(\omega)$. Otherwise, we carry out the above procedure once more. Since we increase the degree of $d(\omega)$ by at least one each time, we shall eventually arrive at (78).

If we now consider in (72) only the polynomials for which

$$a_{n-1}^{(1)} = a_{n-1}^{(2)},\dots,a_{n-[k]+1}^{(1)} = a_{n-[k]+1}^{(2)},$$

then we can conclude, using Dirichlet's pigeonhole principle and Lemma 15, that there exist $m = c(n)H^{\theta+\{k\}}$ polynomials

$$\begin{aligned} &t_i(\omega) = k_i(\omega)d(\omega), && i = 1,\dots,m,\\ &\deg t_i(\omega) \leq n-[k], && H(t_i(\omega)) \leq 2H,\\ &|t_i(\omega)| < c(n)H^{-(n-k)-2n\varepsilon_1}, && \omega\in I, \end{aligned} \tag{80}$$

where among the polynomials $k_i(\omega)$ there are at least two without roots in common. We denote

$$\deg d(\omega) = l, \qquad H(d(\omega)) = H^\lambda.$$

We then conclude from Lemma 1 that

$$\deg k_i(\omega) \leq n-[k]-l, \qquad H(k_i(\omega)) < c(n)H^{1-\lambda}.$$

The polynomials $k_i(\omega)$ are distinct. From this and the fact that the number of polynomials $t_i(\omega)$ is at least $c(n)H^{\theta+\{k\}}$ we obtain, assuming that the $k_i(\omega)$ take small values in I,

$$(1-\lambda)(n-[k]-l) \geq \theta + \{k\} = p_1\left(1-\frac{n+1}{w+1}\right) + \{k\}. \tag{81}$$

It follows from (80) that either $d(\omega)$ or the $k_i(\omega)$ take small values in I. Let us assume that

$$|d(\omega)| < H(d)^{-((w+1)(l+1)/(n+1))+1} < c(n)H^{-((w+1)(l+1)\lambda/(n+1))+\lambda} \tag{82}$$

for some set $B \subset I$, $|B| \geq \frac{1}{2}|I|$. Then from (82) and Lemma 10 we obtain

$$|d(\omega)| < c(n)H^{-((w+1)(l+1)\lambda/(n+1))+\lambda} \tag{83}$$

for all $\omega \in I$. Since $\deg d(\omega) < n$, by the inductive assumption the Hausdorff dimension of the set of points ω belonging to an infinite number of intervals I at all points of which (76) holds does not exceed

$$\frac{l+1}{(w+1)(l+1)/(n+1) - 1 + 1} = \frac{n+1}{w+1}. \tag{84}$$

Therefore, it remains to prove the theorem in the case when ω belongs to an infinite number of intervals I for whose points inequality (82) does not hold in a set B with measure $|B| \geq \frac{1}{2}|I|$. In this case, by (76), the following inequality holds in some set B_1, $|B_1| > \frac{1}{2}|I|$:

$$|k_i(\omega)| < c(n)H^{-(n-k)+\lambda((w+1)(l+1)/(n+1))-\lambda-2n\varepsilon_1}. \tag{85}$$

We again apply Lemma 10. We find that (85) holds for all $\omega \in I$ with a somewhat larger value of $c(n)$. It follows from the definition of the polynomials $k_i(\omega)$ that there exist two polynomials $k'(\omega)$ and $k''(\omega)$ without roots in common. We apply Lemma 12 to them. Here

$$\mu = 1 - \lambda, \qquad s = n - [k] - l = l_2/T + p_1 + \{k\} - l - 1,$$
$$\tau = \frac{l_2}{T} + p_1 - 1 + (n-1)\varepsilon_1 - \frac{w+1}{n+1}(l+1)\lambda + \lambda, \qquad \eta = \frac{l_2}{T} - (n-1)\varepsilon_1.$$

We obtain

$$\frac{l_2}{T} + p_1 + (n-1)\varepsilon_1 - \frac{w+1}{n+1}(l+1)\lambda + 2\max\left(p_1 - \frac{w+1}{n+1}(l+1)\lambda, 0\right) \leq 2(1-\lambda)(l_2/T + p_1 + \{k\} - l - 1). \tag{86}$$

If

$$p_1 < \frac{w+1}{n+1}(l+1)\lambda, \tag{87}$$

then, together with (81), inequalities (86) and (87) form the system of inequalities

$$\begin{gathered} (1-\lambda)\left(\frac{l_2}{T} + p_1 + \{k\} - l - 1\right) \geq p_1\left(1 - \frac{n+1}{w+1}\right) + \{k\}, \\ 2(1-\lambda)\left(\frac{l_2}{T} + p_1 + \{k\} - l - 1\right) \geq \frac{l_2}{T} + p_1 - \frac{w+1}{n+1}(l+1)\lambda + (n-1)\varepsilon_1, \\ p_1 < \frac{w+1}{n+1}(l+1)\lambda. \end{gathered} \tag{88}$$

If

$$p_1 \geq \frac{w+1}{n+1}(l+1)\lambda, \tag{89}$$

then (81), (86), and (89) lead to the system of inequalities

$$\begin{aligned} (1-\lambda)\left(\frac{l_2}{T}+p_1+\{k\}-l-1\right) &\geq p_1\left(1-\frac{n+1}{w+1}\right)+\{k\}, \\ 2(1-\lambda)\left(\frac{l_2}{T}+p_1+\{k\}-l-1\right) &\geq \frac{l_2}{T}+3p_1-3\frac{w+1}{n+1}(l+1)\lambda+(n-1)\varepsilon_1, \\ p_1 &\geq \frac{w+1}{n+1}(l+1)\lambda. \end{aligned} \tag{90}$$

LEMMA 16. *The system of inequalities* (88) *for* $w < 2n+1$ *is contradictory.*

The proof of Lemma 16 requires rather tedious calculations, which as a rule we shall omit. The first and third inequalities of system (88) in the case when the curves

$$(1-\lambda)\left(\frac{l_2}{T}+\{k\}-l-1\right)=p_1\left(1-\frac{n+1}{w+1}\right)+\{k\}, \tag{91}$$

$$\frac{w+1}{n+1}(l+1)\lambda=p_1 \tag{92}$$

intersect define the domain where we shall examine the function

$$f(\lambda,l)=2(1-\lambda)(l_2/T+p_1+\{k\}-l-1)+\frac{w+1}{n+1}(l+1)\lambda.$$

The condition under which the curves (91) and (92) intersect has the form

$$\frac{l_2}{T}\geq 2\sqrt{p_1^2\frac{n+1}{w+1}\left(1-\frac{n+1}{w+1}\right)+p_1\{k\}}. \tag{93}$$

If (93) holds, then at the intersection points of the curves (91) and (92)

$$f(\lambda,l)=2p_1\left(1-\frac{n+1}{w+1}\right)+p_1+2\{k\}.$$

But if (93) holds and $w < 2n+1$, then

$$\frac{l_2}{T}\geq 2p_1\left(1-\frac{n+1}{w+1}\right)+2\{k\}$$

and so at the intersection points system (88) is contradictory. System (88) also turns out to be contradictory on the boundaries (91) and (92) of the domain. A stationary point of the function $f(\lambda,l)$ cannot lie inside the domain if at this point

$$f(\lambda,l)<l_2/T+p_1+(n-1)\varepsilon_1.$$

This follows from the fact that otherwise there would exist a closed curve situated entirely inside the domain and containing a stationary point. But the curve

$$f(\lambda,l)=l_2/T+p_1+(n-1)\varepsilon_1$$

is not closed.

LEMMA 17. *The system of inequalities* (90) *for* $w < 2n+1$ *is contradictory.*

The proof is similar to the proof of Lemma 16.

PROPOSITION 3. *For* $n < w < 2n+1$

$$\dim \mathscr{L}_n(w) \leq (n+1)/(w+1).$$

§7. The degenerate case for $w > 4n+3$

The degenerate case for $w \geq 2n+1$ requires a more complicated construction. We shall begin its examination with common constructions both for the case $2n+1 \leq w \leq 4n+3$ and for the case $w > 4n+3$.

We denote by ρ_r a number of the form r/T, where r is a nonnegative integer. If for any r the number of intervals I to each of which belong $H^{k+\theta-1+\rho_r+\kappa\varepsilon_1}$ polynomials, $0 \leq \kappa \leq 1$, does not exceed $H^{n-p_1-k+1-\rho_r}$, then the theorem can be proved without difficulty, since in this case the total number of polynomials with the prescribed p_1 does not exceed $c(n)H^{n-p_1+\theta+\varepsilon_1}$, and the series

$$\sum_{H=1}^{\infty} H^{n-p_1+\theta+\varepsilon_1} H^{(-w+p_1-1)((n+1)/(w+1)+\varepsilon)}$$

converges.

Therefore, in what follows we assume that there exists at least one integer r for which there are at least $H^{n-k-p_1+1-\rho_r}$ intervals I to each of which belong $H^{k+\theta-1+\rho_r+\kappa\varepsilon_1}$ polynomials. We fix one such r, and in what follows we denote ρ_r by ρ. We denote by $\mathscr{I}(\rho)$ the set of intervals I to each of which belong $H^{k+\theta-1+\rho+\kappa\varepsilon_1}$ polynomials. We shall again use Dirichlet's pigeonhole principle and for each $I \in \mathscr{I}(\rho)$ shall construct the set of polynomials $\Psi(I)$ consisting of polynomials

$$t_1(\omega), \dots, t_m(\omega), \qquad m > c(n)H^{\theta+\rho+\{k\}},$$

each of which satisfies, for all $\omega \in I$, the inequality

$$|t_j(\omega)| < c(n)H^{-(n-k)-2n\varepsilon_1}, \qquad 1 \leq j \leq m, \tag{94}$$

where for all j, according to (72) and (73),

$$H(t_j) \leq 2H^{1-i/(n-[k]+i)}, \qquad \deg t_j \leq n-[k]+i.$$

It is not hard to see that the polynomial $P_0(x)$ can be so chosen that for the first $c(n)H^{\{k\}+\{\theta\}}$ polynomials the coefficients of $x^{n-[k]+i}, \dots, x^{n-[k]-[\theta]+1}$ coincide. We would then obtain

$$H(t_1) \leq 2H^{1-i/(n-[k]+i)}, \qquad \deg t_1 \leq n-[k]-[\theta]+i. \tag{95}$$

If among the polynomials $t_j(\omega)$ there exists at least one having no roots in common with $t_1(\omega)$, then, using Lemma 12 and the inequality $\theta \geq 1$, we obtain a contradiction. In the case when there are no such polynomials, by Lemma 15 we can pass to a subset $\Psi_1(I) \subset \Psi(I)$ consisting of polynomials

$$t_1(\omega), \dots, t_{im_1}(\omega), \qquad m_1 > c(n)H^{\theta+\rho+\{k\}},$$

representable in the form

$$t_j(\omega) = k_j(\omega)d(\omega). \tag{96}$$

We note that among the polynomials $k_j(\omega)$, $j \geq 2$, there exists a polynomial relatively prime to $k_1(\omega)$. Let $I \in \mathscr{I}(\rho)$. We denote by a_1 a rational number with denominator T such that the inequality

$$|d(\omega)| < H^{-(n-k)+a_1-n\varepsilon_1}$$

holds for some set $B \subset I$, $|B| \geq \frac{1}{2}|I|$, but the inequality

$$|d(\omega)| < H^{-(n-k)-a_1-(n+1)\varepsilon_1} \tag{97}$$

holds only for some set $B_1 \subset B$, $|B_1| \leq \frac{1}{2}|I|$. Since the inequality opposite to (97) holds for the set $B_2 = I \setminus B_1$, from (94) we obtain

$$|k_j(\omega)| < c(n)H^{-a_1-(n-1)\varepsilon_1} \tag{98}$$

for all $\omega \in B_2$, $|B_2| \geq \frac{1}{2}|I|$.

Let us pass from the set of polynomials $\Psi_1(I)$ to the set of polynomials $\Psi_2(I)$ consisting of the polynomials $t_{i1}(\omega), \dots, t_{im_1}(\omega)$; i.e., we exclude $t_1(\omega)$ from $\Psi_2(I)$. All polynomials in $\Psi_2(I)$ are representable in the form (96); however, among the polynomials $k_{i1}(\omega), \dots, k_{im_1}(\omega)$ it might turn out that there are no polynomials that are relatively prime. We apply Lemma 15. We find that the polynomials from some subset $\Psi_3(I) \subset \Psi_2(I)$ are representable in the form

$$t_j(\omega) = k_j(\omega)d_1(\omega), \qquad j = 1, \dots, m_2, \qquad m_2 > c(n)H^{\theta+\rho+\{k\}} \tag{99}$$

and among the polynomials $k_j(\omega) \in \Psi_3(I)$ there exist polynomials without roots in common. From (96) and (99) we can conclude that $d(\omega)|d_1(\omega)$, whence, using Lemma 1, we get

$$\deg d \leq \deg d_1, \qquad H(d) < c(n)H(d_1). \tag{100}$$

We denote by a a rational number with denominator T such that the inequality

$$|d_1(\omega)| < H^{-(n-k)+a-n\varepsilon_1}$$

holds for some set with measure at least $\frac{1}{2}|I|$, but the inequality

$$|d_1(\omega)| < H^{-(n-k)+a-(n+1)\varepsilon_1} \tag{101}$$

holds only for some set with measure at most $\frac{1}{2}|I|$. Then, by analogy with (98), we obtain

$$|k_j(\omega)| < c(n)H^{-a-(n-1)\varepsilon_1} \tag{102}$$

for some set with measure at least $\frac{1}{2}|I|$ and any $k_j(\omega) \in \Psi_3(I)$.

It is not hard to see that for each $j = 1, \dots, m_2$ there exists a rational number $\sigma = m/T \geq 0$ such that (102) holds for all ω in some interval I', $|I'| = H^{-l_2/T+\sigma}$, with center at the point $O_{T'}$ but does not hold for any subset of the interval $[O_{I'} - H^{-l_2/T+\sigma+\varepsilon_1}, O_{I'} + H^{-l_2/T+\sigma+\varepsilon_1}]$ with measure greater than $H^{-l_2/T+\sigma+\varepsilon_1}$. Since σ can assume a finite number of values that depends on n and ε, we shall further assume that for some σ there exist at least $c(n)H^{\rho+\sigma}$

polynomials $k_j(\omega)$ with the same σ. We fix one such value of σ, and we denote by $\Psi_4(I)$ the set of polynomials $t_j(\omega)$ with the same σ.

Since $\Psi_4(I) \subset \Psi_3(I)$, a decomposition into factors of the form (99) is valid for any $t_j(\omega)$. However, among the polynomials $k_j(\omega)$ for $t_j(\omega) \subset \Psi_4(I)$ it is no longer possible to find two without roots in common. Then one should again apply Lemma 15 and pass to a set $\Psi_5(I)$ consisting of at least $c(n)H^{\theta+\rho}$ polynomials $t_j(\omega)$ representable in the form $t_j(\omega) = k_j(\omega)d_2(\omega)$. We then again obtain inequalities for the polynomials $d_2(\omega)$ and $k_j(\omega)$ analogous to (101) and (102). Since we obtain an increase of at least one in the degree of the polynomials $d_i(\omega)$ each time, we eventually obtain a set $\Psi_5(I)$ consisting of at least $c(n)H^{\theta+\rho+\{k\}}$ polynomials, where all the $t_j(\omega) \in \Psi_s(I)$ have the form $t_j(\omega) = k_j(\omega)d_{j(s)}(\omega)$ and among the polynomials $k_j(\omega)$ there are at least two without roots in common. So as not to change the notation, we shall assume that we have stopped at the first step, and so $\Psi_s(I) = \Psi_4(I)$.

We set

$$H(d_1) = H^{\lambda_1}, \qquad \deg d_1 = l_1 \tag{103}$$

We further assume that the set $\mathscr{I}(\rho)$ contains at least $c(n)H^{l_2/T-\rho}$ intervals I such that the polynomials $k_j(\omega)$, $d(\omega)$, and $d_1(\omega)$ constructed for distinct intervals I have the same degree and heights lying between rational powers of H of the form p/T and $(p+1)/T$. We shall also assume that the exponents a in inequalities of the form (102) and the numbers σ are identical. This can be explained by the fact that the number of distinct possibilities of powers for a and σ and rational values of the exponents for the heights of $k_j(\omega)$, $d_1(\omega)$, and $d(\omega)$ is finite and depends on n and ε.

Let us construct a set χ from the polynomials $d_1(\omega)$ constructed for distinct intervals I. It consists of at least $c(n)H^{l_2/T-\rho}$ not necessarily distinct polynomials. Clearly there exists a rational number $\gamma = s/T$ such that the number of coincident polynomials in the set χ is included between $H^{s/T}$ and $H^{(s+1)/T}$, and the number of distinct polynomials in χ is at least $c(n)H^{l_2/T-\rho-\gamma-\varepsilon_1}$. We denote by $\mathscr{I}_\gamma(\rho)$ the set of intervals $I \subset \mathscr{I}(\rho)$ with the same polynomial $d_1(\omega)$.

Let

$$\chi_1 = \bigcup_{I\in\mathscr{I}_\gamma(\rho)} \bigcup_{j=1}^{m_2} k_j(\omega).$$

Then the set χ_1 consists of at least $c(n)H^{\theta+\rho+\gamma+\{k\}}$ polynomials. However, each of the polynomials $k_j(\omega)$ can occur in $H^{\sigma+\kappa_1\varepsilon_1}$, $0 \le \kappa_1 \le 1$, intervals. We conclude from this that the number of distinct polynomials $k_j(\omega)$ in the set χ_1 is at least $c(n)H^{\theta+\rho+\gamma+\{k\}-\sigma-\varepsilon_1}$. For the polynomials $k_j(\omega) \in \Psi_4(I)$ inequality (102) holds in a set with measure at least $H^{-l_2/T}$. Since the set $\Psi_4(I)$ contains polynomials without roots in common, we have, applying Lemma 12,

$$\begin{aligned}\left(1 - \frac{i}{n-[k]+i} - \lambda_i\right)&(n-[k]+i-l_i)\\ &\ge a+1-\frac{i}{n-[k]+i} + 2\max\left(a+1-\frac{i}{n-[k]+i} - \frac{l_2}{T} + \sigma, 0\right).\end{aligned} \tag{104}$$

Since the number of distinct polynomials $d_1(\omega)$ is at least $c(n)H^{l_2/T-\rho-\gamma-\varepsilon_1}$ and the number of distinct polynomials $k_j(\omega)$ is at least $c(n)H^{\theta+\rho+\gamma+\{k\}-\sigma-\varepsilon_1}$, by Lemma 9 we have

$$\lambda_i(l_i+1) \geq l_2/T - \rho - \gamma - \varepsilon_1, \tag{105}$$

$$\left(1 - \frac{i}{n-[k]+i} - \lambda_i\right)(n-[k]+i-l_i+\max(0,-a-(n-1)\varepsilon_1) \\ \geq \theta+\rho+\gamma+\{k\}-\sigma-\varepsilon_1.$$

We shall assume in what follows that $\max(0,-a-(n-1)\varepsilon_1)=0$, since otherwise it will be easier for us to show that the exponent $n-k-a$ is not small, and this is indeed the goal of our subsequent investigations. In this case the second inequality in (105) assumes the form

$$\left(1 - \frac{i}{n-[k]+i} - \lambda_i\right)(n-[k]+i-l_i) \geq \theta+\rho+\gamma-\sigma-\varepsilon_1,$$

which, together with (104) and (105), we write as the system of inequalities

$$\left(1 - \frac{i}{n-[k]+i} - \lambda_i\right)(n-[k]+i-l_i) \geq \theta+\rho+\gamma-\sigma+\{k\}-\varepsilon_1,$$

$$2\left(1 - \frac{i}{n-[k]+i} - \lambda_i\right)(n-[k]+i-l_i) \geq a+1-\frac{i}{n-[k]+i}-\lambda_i l_i \\ \geq \frac{l_2}{T}-\rho-\gamma-\varepsilon_1-\lambda_i+2\max\left(a+1-\frac{i}{n-[k]+i}-\lambda_i-\frac{l_2}{T}+\sigma,0\right). \tag{106}$$

We set $\tau=\gamma+\rho$ and shall assume further that

$$l_2/T = p_1. \tag{107}$$

To simplify the notation, in what follows we shall omit the ε_1 on the right sides of the inequalities. Then the system (106) assumes the form

$$\left(1 - \frac{i}{n-[k]+i} - \lambda_i\right)(n-[k]+i-l_i) \geq \theta+\tau-\sigma+\{k\},$$

$$2\left(1 - \frac{i}{n-[k]+i} - \lambda_i\right)(n-[k]+i-l_i) \geq a+1-\frac{i}{n-[k]+i} \\ -\lambda_i+\lambda_i l_i \geq p_1-\tau+2\max\left(a+1-\frac{i}{n-[k]+i}-\lambda_i-p_1+\sigma,0\right). \tag{108}$$

LEMMA 18. *The system of inequalities*

$$\left(1 - \frac{i}{n-[k]+i} - \lambda_i\right)(n-[k]+i-l_i) > (1-\sqrt{c})^2(n-k), \qquad \lambda_i l_i > c(n-k)$$

is contradictory for any i, λ_i, and l_i with $0 \leq i \leq [k]$, $0 \leq \lambda_i \leq 1-i/(n-[k]+i)$, $0 \leq l_i \leq n-[k]+i$, and $0<c<1$.

PROOF. The function

$$f(\lambda_i,l_i) = \left(\frac{n-k}{n-[k]+i} - \lambda_i\right)(n-[k]+i-l_i)$$

can assume its largest value either on the boundary of the domain defined by the system of inequalities

$$0 \le \lambda_i \le \frac{n-k}{n-[k]+i}, \qquad 0 \le l_i \le n-[k]+i, \qquad \lambda_i l_i \ge c(n-k)$$

or at a stationary point. But a stationary point of the function $f(\lambda_i, l_i)$ lies on the boundary of the domain. On the lines $\lambda_i = (n-k)/(n-[k]+i)$ and $l_i = n-[k]+i$ we have $f(\lambda_i, l_i) = 0$, and on the curve $\lambda_i l_i = c(n-k)$ we have

$$f(\lambda_i, l_i) \le (1-\sqrt{c})^2(n-k).$$

We introduce the notation

$$\mu_i = 1 - \frac{i}{n-[k]+i}, \qquad n_i = n-[k]+i, \qquad p = 2p_1 - 1. \tag{109}$$

Since in (108) we must obtain the best possible upper bound for a, in (108) we set $\{k\} = 0$.

Since $w > 4n+3$, we conclude from Lemma 18 that the system of inequalities

$$(\mu_i - \lambda_i)(n_i - l_i) > 0.75p_1 - 0.375, \qquad \lambda_i l_i > 0.16p$$

is contradictory. Therefore, in what follows we shall consider only the $\lambda_i l_i$ that lie in the interval $[0, 0.16p]$. We shall divide this interval into the disjoint intervals

$$(0.1225p, 0.16p], (0.09p, 0.1225p], (0.0625p, 0.09p], (0.04p, 0.0625p],$$

$$(0.0225p, 0.04p], (0.01p, 0.0225p], [0, 0.01p].$$

Let $\lambda_i l_i \in (0.1225p, 0.16p]$. Then from the third inequality in (108) we obtain $\tau > 0.34p + 0.5$. By Lemma 18 the system

$$(\mu_i - \lambda_i)(n_i - l_i) > 0.4225p, \qquad \lambda_i l_i > 0.1225p$$

is contradictory. Therefore, if $\sigma < 0.2925p + 0.875$, then the first inequality in (108) assumes the form $(\mu_i - \lambda_i)(n_i - l_i) > 0.4225p$, and system (108) with the condition $\lambda_i l_i \in (0.1225p, 0.16p]$ becomes contradictory. If $\sigma \ge 0.2925p + 0.875$, then from the second inequality in (108) we obtain $3a + 3(\mu_i - \lambda_i) - 0.415p + 0.75 < 0.845p$, since the reverse inequality would again lead to a contradiction. We further have $a \le 0.42p + (-\mu_i + \lambda_i) - 0.25$, whence $n - k - a \ge 0.58p + \mu_i - \lambda_i + 0.25$. Let us use Lemma 14. We find that there exists $d'_{1i}(\omega)$, a divisor of $d_{1i}(\omega)$, having degree at most l_i and height at most $c(n)H^{\lambda_i}$, that is either an irreducible polynomial or a power of an irreducible polynomial and such that

$$|d'_{1i}(\omega)| < c(n)H^{-0.42p-0.25} \tag{110}$$

for all ω in some union of intervals I. Since the exponent in (110) is less than 0, using Lemma 9 one can replace $\lambda_i(l_i+1)$ by $\lambda_i l_i$ in the first inequality in (105).

Let us now obtain upper bounds for the height of the polynomial $d'_{1i}(\omega)$ and its degree. Here we shall assume that $(\mu_i - \lambda_i)(n_i - l_i)$ belongs to a sufficiently small neighborhood of $0.4225p$. If $(\mu_i - \lambda_i)(n_i - l_i)$ assumes different values of the interval $[0.375p, 0.4225p]$, then in this case it would be necessary to make finer divisions of the interval $[0, 0.16p]$. We shall, for example, assume that $(\mu_i - \lambda_i)(n_i - l_i) > 0.41p$. Since $\lambda_i \geq 0$, we obtain $l_i < 0.59(p+i)$, whence it follows from $\lambda_i l_i > 0.1225p$ that $\lambda_i > 0.1225p/(0.59(p+i))$. Now from the inequality $0.8\mu_i(n_i - l_i) > 0.41p$ we obtain $l_i < 0.5n_i$, whence $\lambda_i > 0.24\mu_i$, and so forth. In several steps we obtain $l_i < 0.4n_i$. We carry out a similar argument for λ_i. We obtain $\lambda_i < 0.4\mu_i$.

Thus, we have obtained the existence of a polynomial $d'_{1i}(\omega)$ that satisfies (110). Moreover, the degree and height of this polynomial satisfy the inequalities

$$H(d'_{1i}) < c(n)H^{0.4\mu_i}, \qquad \deg d'_{1i} < 0.4n_i.$$

If $\lambda_i l_i \in (0.09p, 0.1225p]$ or any other interval up to $(0.01p, 0.0225p]$, then one can carry out similar arguments. Moreover, we obtain the following results:

If $\lambda_i l_i \in (0.09p, 0.1225p]$, then

$$|d'_{1i}(\omega)| < H^{-0.39p-0.25}, \qquad l_i < 0.35n_i, \qquad \lambda_i < 0.35\mu_i. \tag{111}$$

If $\lambda_i l_i \in (0.0625p, 0.09p]$, then

$$|d'_{1i}(\omega)| < H^{-0.37p-0.25}, \qquad l_i < 0.3n_i, \qquad \lambda_i < 0.3\mu_i. \tag{112}$$

If $\lambda_i l_i \in (0.04p, 0.0625p]$, then

$$|d'_{1i}(\omega)| < H^{-0.29p-0.25}, \qquad l_i < 0.25n_i, \qquad \lambda_i < 0.25\mu_i. \tag{113}$$

If $\lambda_i l_i \in (0.0225p, 0.04p]$, then

$$|d'_{1i}(\omega)| < H^{-0.21p-0.25}, \qquad l_i < 0.2n_i, \qquad \lambda_i < 0.2\mu_i. \tag{114}$$

If $\lambda_i l_i \in (0.01p, 0.0225p]$, then

$$|d'_{1i}(\omega)| < H^{-0.13p-0.25}, \qquad l_i < 0.13n_i, \qquad \lambda_i < 0.13\mu_i. \tag{115}$$

If $\lambda_i l_i < 0.01p$, then we pass from the set of polynomials $\Psi_2(I)$ to the set $\Psi_1(I)$. We apply Lemma 12 to the polynomial $k_1(\omega)$ and any one in $\Psi_1(I)$ relatively prime to it. Bearing in mind that the degree of $t_1(\omega)$ does not exceed $0.625p + \{\theta\} + \{k\} + i$, we obtain

$$\begin{aligned}(\mu_i - \lambda_i)(1.625p + 2i + 2\{k\} - \{\theta\} - 2l_i)& \\ \geq a + \mu_i - \lambda_i + 2\max(a + \mu_i - \lambda_i - p_1, 0),&\end{aligned} \tag{116}$$

$$0 \leq \lambda_i \leq \mu_i, \qquad 1 \leq l_i \leq 0.625p + i + \{k\} + \{\theta\}$$

By analogy with the way it was done in obtaining (110)–(115), one can obtain from (116) the existence of polynomials $d'_{1i}(\omega)$ that have either small degree or height and assume small values for all $\omega \in I$. Since this procedure is similar to (110)–(115), we shall omit it.

Let us now show that the violation of condition (107), as a rule, leads to an increase in the exponents in (110)–(115). But if an exponent decreases, then it is by an insignificant value. Let $l_2/T < p_1$. Let us return to the system (106) by setting $\delta = p_1 - l_2/T$. From the definitions of l_2/T, δ, and p_1 one can obtain $0 \le \delta \le (n-1)p_1/n$. We again set $\{k\} = 0$ and omit ε_1. We obtain

$$(\mu_i - \lambda_i)(n_i - l_i - \delta) > \theta + \tau - \sigma,$$

$$\begin{aligned} 2(\mu_i - \lambda_i)(n_i - l_i - \delta) \\ \ge a + \mu_i - \lambda_i + 2\max(a + \mu_i - \lambda_i - 0.5p - 0.5 + \delta + \sigma, 0), \end{aligned} \tag{117}$$

$$\lambda_i l_i \ge p_1 - \tau - \sigma.$$

By performing simple transformations we write (117) in a form close to (108):

$$\begin{aligned} (\mu_i - \lambda_i)(n_i - l_i) &\ge \theta + \tau - \sigma + \delta(\mu_i - \lambda_i), \\ 2(\mu_i - \lambda_i)(n_i - l_i) &\ge a + \mu_i - \lambda_i + 2(\mu_i - \lambda_i)\delta \\ &\quad + 2\max(a + \mu_i - \lambda_i - 0.5p - 0.5 + \delta + \sigma, 0), \\ \lambda_i l_i &\ge p_1 - \tau - \delta. \end{aligned} \tag{118}$$

We again divide the interval $[0, 0.16p)$ into the same seven disjoint intervals and carry out arguments similar to those used in obtaining (110)–(115). We find that if the exponents in (110)–(112) in fact change, then they remain at most

$$-0.41p - 0.25; \qquad 0.38p - 0.25; \qquad -0.36p - 0.25 \tag{119}$$

respectively.

Let us prove that all polynomials $d'_{1i}(\omega)$ $(i = 0, \dots, [k])$ can only be powers of the same polynomial. We consider the polynomials $d'_{1i}(\omega)$ and $d'_{1,i+1}(\omega)$. If they are not powers of the same polynomial, then one can apply Lemma 12 to them. Moreover, if

$$\lambda_i l_i \in (0.1225p, 0.16p] \quad \text{and} \quad \lambda_{i+1} l_{i+1} \in (0.09p, 0.1225p],$$

then

$$\begin{aligned} H^{\lambda_i l_{i+1} + \lambda_{i+1} l_i} &< H^{0.4(p/(p+i))\cdot 0.35(p+i+1) + 0.35(p/(p+i+1))\cdot 0.4(p+i)} \\ &< H^{0.14p((p+i+1)/(p+i) + (p+i)/(p+i+1))}. \end{aligned} \tag{120}$$

But

$$\frac{p+i+1}{p+i} + \frac{p+i}{p+i+1} < 2 + \frac{1}{p(p+1)}$$

for all i. Since $p \ge 2$, the right side of (119) can be bounded from above by the value $H^{0.28p+0.5}$.

Bearing (110)–(115) and (119) in mind, one can rewrite (120) in the form

$$\lambda_i l_i \in (0.1225p, 0.16p], \qquad \lambda_{i+1} l_{i+1} \in (0.09p, 0.1225p],$$

and

$$\begin{aligned} H^{\lambda_i l_{i+1} + \lambda_{i+1} l_i} &< H^{0.28p+0.05}, \\ \max(|d'_{1i}(\omega)|, |d'_{1,i+1}(\omega)|) &< H^{-0.37p-0.25}. \end{aligned} \tag{121}$$

Analyzing the remaining possibilities for $\lambda_i l_i$ and $\lambda_{i+1} l_{i+1}$, we see that if $d'_{1i}(\omega)$ and $d_{1,i+1}(\omega)$ are not powers of the same irreducible polynomial, then by Lemma 12 we obtain a contradiction.

Thus, there exists an irreducible polynomial $d_0(\omega)$ such that $d'_{1,i}(\omega) = d_0^{s_i}(\omega)$. Let us consider $i = 0$ and $i = [k] - 1$. From the fact that $\lambda_i l_i$ belongs to one of the six chosen intervals we obtain

$$|d_0^{s_0}(\omega)| < H^{-\nu_j p - 0.25}, \qquad \deg d_0^{s_0}(\omega) < \tau_j p, \qquad j = 1, \dots, 6, \tag{122}$$

where ν_j and τ_j depend on the number of the chosen intervals and are defined from (110)–(115) and (119). From these same inequalities and (122) we have

$$|d_0(\omega)| < H^{-(\nu_j p + 0.25)/s_0}, \qquad \deg d_0 \le \tau_j p / s_0, \tag{123}$$

$$\begin{aligned} &\nu_1 = 0.41; \quad \nu_2 = 0.38; \quad \nu_3 = 0.36; \quad \nu_4 = 0.29; \quad \nu_5 = 0.21; \quad \nu_6 = 0.13, \\ &\tau_1 = 0.4; \quad \tau_2 = 0.35; \quad \tau_3 = 0.3; \quad \tau_4 = 0.25; \quad \tau_5 = 0.2; \quad \tau_6 = 0.13. \end{aligned}$$

$$|d_0(\omega)| < H^{-(\nu_{j'} p + 0.25)/s_k}, \qquad H(d_0) < c(n) H^{(\tau_{j'} p)/(s_k n)}. \tag{124}$$

Comparing (123) and (124), we obtain

$$\begin{aligned} &|d_0(\omega)| < H^{-\max((\nu_j p + 0.25)/s_0,\ (\nu_{j'} p + 0.25)/s_k)}, \\ &H(d_0) < c(n) H^{(\tau_{j'} p)/(s_k n)}, \qquad \deg d_0 < \tau_j p / s_0. \end{aligned} \tag{125}$$

The polynomial $d_0(\omega)$ has no roots in common with a polynomial $P(\omega)$ belonging to I. We apply Lemma 13 to these polynomials. We obtain

$$\frac{\tau_j}{s_0} p \cdot 1 + \frac{\tau_{j'}}{s_k} \cdot \frac{p}{n} \cdot n < 2 \max\left(\frac{\nu_j p}{s_0}, \frac{\nu_{j'} p}{s_k} \right),$$

since $\nu_j \ge \tau_j$ and $\nu_{j'} \ge \tau_{j'}$. Thus, we have proved the following proposition.

PROPOSITION 4. $\dim \mathscr{L}_n(w) \le (n+1)/(w+1)$ *for* $w \ge 4n + 3$.

§8. The degenerate case for $2n + 1 \le w \le 4n + 3$

The arguments of the preceding section were based on the fact that we obtained a good approximation of a zero in intervals I by values of polynomials whose degree and height are relatively small. We now note further that if $|d(\omega)| < H^{-(n-k)+a}$, then, since $d(\omega)$ has degree less than n, the following inequality must hold:

$$\frac{l+1}{\frac{n-k-a}{\lambda} + 1} > \frac{n+1}{w+1}, \tag{126}$$

because, otherwise, everything is proved. We rewrite (126) in the form

$$\lambda(l+1) > \frac{n+1}{w+1}(n - k - a + \lambda). \tag{127}$$

Inequality (127), together with the system (108) taken for $i = 0$ and $\{k\} = 0$, forms the system

$$\begin{aligned} (1-\lambda_0)(p-l_0) &\geq \theta + \tau - \sigma, \\ 2(1-\lambda_0)(p-l_0) &\geq a + 1 - \lambda_0 + 2\max(a+1-\lambda_0 - p_1 + \sigma, 0), \\ \lambda_0 l_0 &\geq p_1 - \tau = 0.5p + 0.5 + \tau, \\ \lambda_0(l_0+1) &> \frac{n+1}{w+1}(p - a + \lambda_0). \end{aligned} \tag{128}$$

Since $\theta \geq 0.5p_1$, we have $\lambda_0 l_0 \leq 0.25p$, because otherwise, by Lemma 18, we find that the system (128) is contradictory. We divide the interval $[0, 0.25p)$ into the disjoint intervals

$$[0.2209p, 0.25p), \qquad [0.2205p, 0.2209p), \qquad [0.16p, 0.2025p)$$

and proceed as in the case $w > 4n + 3$.

Let $\lambda_0 l_0 \in [0.2209p, 0.25p)$. Then from the third inequality in (128) we obtain $\tau > 0.25p + 0.5$. If now $\sigma \leq 0.2191p + 0.75$, then by Lemma 18 the first and third inequalities in (128) are contradictory. We therefore assume that $\sigma > 0.2191p + 0.75$. Then from the second inequality in (128) we conclude that $a < 0.38p - 1.16 + \lambda_0$, whence $p - a + \lambda_0 \geq 0.62p + 1.16$. For $2n + 1 \leq w \leq 2.3n + 1.3$, bearing in mind that $p \geq 3$, we find that the fourth inequality in (128) is contradictory.

If $w \geq 2.3n + 1.3$, then $\theta > 0.281p$ and by Lemma 18 we obtain a contradiction. We shall now assume that $\lambda_0 l_0 \in [0.2025p, 0.2209p)$. We again carry out this procedure. We again find that system (128) is contradictory. A contradiction is also obtained for $\lambda_0 l_0$ belonging to different intervals. We have proved the following proposition.

PROPOSITION 5. *For* $2n + 1 \leq w \leq 4n + 3$

$$\dim \mathscr{L}_n(w) \leq \frac{n+1}{w+1}.$$

Thus, the proof of the theorem is complete.

REFERENCES

1. V. I. Bernik, *A metric theorem on the simultaneous approximation of a zero by the values of integral polynomials*, Izv. Akad. Nauk SSSR Ser. Mat. **44** (1980), 24–45; English transl. in Math. USSR Izv. **16** (1981).

2. ——, *On the Baker–Schmidt conjecture*, Dokl. Akad. Nauk BSSR **23** (1979), 392–395. (Russian)

3. A. O. Gel'fond, *Transcendental and algebraic numbers*, GITTL, Moscow, 1952; English transl., Dover, New York, 1960.

4. J. W. S. Cassels, *An introduction to Diophantine approximation*, Cambridge Univ. Press, 1957.

5. Yu. V. Nesterenko, *A lower bound for a collection of polynomials*, All-Union School on Number Theory, Abstracts of Reports, Dushanbe, 1977. (Russian)

6. V. G. Sprindzhuk, *On a conjecture of Mahler*, Dokl. Akad. Nauk SSSR **154** (1964), 783–786; English transl. in Soviet Math. Dokl. **5** (1964).

7. ——, *Further remarks on a conjecture of Mahler*, Dokl. Akad. Nauk SSSR **155** (1964), 54–56; English transl. in Soviet Math. Dokl. **5** (1964).

8. ——, *Proof of Mahler's conjecture on the measure of the set of S-numbers*, Izv. Akad. Nauk SSSR Ser. Mat. **29** (1965), 379–436; English transl. in Amer. Math. Soc. Transl. (2) **51** (1966).

9. ——, *Mahler's problem in metric number theory*, "Nauka i Tekhnika", Minsk, 1967; English transl., Amer. Math. Soc., Providence, R.I., 1969.

10. N. I. Fel'dman, *Approximation of certain transcendental numbers*. I, Izv. Akad. Nauk SSSR Ser. Mat. **15** (1951), 53–74; English transl. in Amer. Math. Soc. Transl. (2) **59** (1966).

11. A. Baker and Wolfgang M. Schmidt, *Diophantine approximation and Hausdorff dimension*, Proc. London Math. Soc. (3) **21** (1970), 1–11.

12. A. Baker, *On a theorem of Sprindzuk*, Proc. Roy. Soc. London Ser. A **292** (1968), 92–104.

13. R. C. Baker, *Sprindzuk's theorem and Hausdorff dimension*, Mathematika **23** (1976), 184–197.

14. A. S. Besicovitch, *On linear sets of points of fractional dimension*, Math. Ann. **10** (1929), 161–193.

15. R. Güting, *On Mahler's function θ_1*, Michigan Math. J. **10** (1963), 161–179.

16. Felix Hausdorff, *Dimension und äusseres Mass*, Math. Ann. **79** (1918/19), 157–179.

17. Vojtěch Jarník, *Diophantische Approximationen und Hausdorffsches Mass*, Mat. Sb. **36** (1929), 371–382.

18. Friedrich Kasch and Bodo Volkmann, *Zur Mahlerschen Vermutung über S-Zahlen*, Math. Ann. **136** (1958), 442–453.

19. A. Khintchine, *Einige Sätze über Kettenbrüche, mit Anwendungen auf die Theorie der Diophantischen Approximationen*, Math. Ann. **92** (1924), 115–125.

20. Eduard Wirsing, *Approximation mit algebraischen Zahlen beschränkten Grades*, J. Reine Angew. Math. **206** (1960), 67–77.

Translated by R. LENET

Amer. Math. Soc. Transl.
(2) Vol. **140**, 1988

On the Irrationality of Some Quantities Containing $\varsigma(3)$

L. A. GUTNIK

§1. Introduction

In Autumn of 1978, Apéry made a breakthrough in an old question concerning the values $\varsigma(2k+1)$, $k \in \mathbf{N}$, by proving that $\varsigma(3)$ is irrational.

The following result is proved in the present article.

THEOREM 1. *Let $q \in \mathbf{Q}$, $q \neq 0$. Then at least one of the numbers*

$$-\frac{3}{q}\varsigma(3) + \varsigma(2), \qquad \varsigma(2) - 2q\log 2 \tag{1}$$

is irrational.

By supposing that one of the numbers in (1) is zero for some $q \in \mathbf{Q}$, $q \neq 0$, we can arrive at the following result.

THEOREM 2. *There is at least one irrational number in each of the following sets*:

$$\left\{\frac{\varsigma(3)}{\varsigma(2)}, \varsigma(2) - \frac{6\varsigma(3)\log 2}{\varsigma(2)}\right\}, \qquad \left\{\frac{\log 2}{\varsigma(2)}, \varsigma(2) - \frac{6\varsigma(3)\log 2}{\varsigma(2)}\right\}. \tag{2}$$

Here we shall work with the following functions, which belong to a class that was introduced and studied in a series of papers by K. S. Meijer.

In the first place, we consider the function

$$\begin{aligned} f_1(z;\nu) &= -\,C^{1,2}_{4,4}\left(-z\left|\begin{matrix} -\nu, & -\nu, & \nu+1, & \nu+1 \\ 0, & 0, & 0, & 0 \end{matrix}\right.\right) \\ &= -\frac{1}{2\pi i}\int_{L_1} \frac{\Gamma(-s)\Gamma^2(1+\nu+s)}{\Gamma^3(1+s)\Gamma^2(1+\nu-s)}(-z)^s\,ds, \qquad \nu \in \mathbf{N}\cup\{0\}, \end{aligned} \tag{3}$$

where the curve L_1 goes from $+\infty$ to $+\infty$, passing around the set $\mathbf{N}\cup\{0\}$ in the negative direction but not including any point of the set $-\mathbf{N}$. The function

---

1980 *Mathematics Subject Classification* (1985 *Revision*). Primary 11J72; Secondary 11M06.
Translation of Acta Arith. **42** (1983), 255–264; MR **85j**: 11079.

$f_1(z;\nu)$ can be represented as a finite hypergeometric series

$$f_1(z;\nu) = {}_4F_3\left(\begin{matrix}-\nu, -\nu, \nu+1, \nu+1; z\\ 1,1,1\end{matrix}\right) = \sum_{k=0}^{\nu}\binom{\nu}{k}^2\binom{\nu+k}{k}^2 z^k. \tag{4}$$

The value $f_1(1;\nu)$ appeared in Apéry's work.

In the second place, we shall work with the function

$$\begin{aligned} f_2(z;\nu) &= -\,C_{4,4}^{3,2}\left(-z\,\middle|\,\begin{matrix}-\nu, & -\nu, & \nu+1, & \nu+1\\ 0, & 0, & 0, & 0\end{matrix}\right) \\ &= \frac{-1}{2\pi i}\int_{L_2}\frac{\Gamma^3(-s)\Gamma^2(1+\nu+s)}{\Gamma(1+s)\Gamma^2(1+\nu-s)}(-z)^s\,ds, \qquad \nu\in\mathbf{N}\cup\{0\}, \end{aligned} \tag{5}$$

where the curve L_2 goes from $-\infty$ to $-\infty$, passing around the set $-\mathbf{N}$ in the positive direction but not including any point of the set $\mathbf{N}\cup\{0\}$. We have the following expansion for $f_2(z;\nu)$:

$$f_2(z;\nu) = \sum_{k=0}^{\infty}\left(\prod_{\kappa=k+1}^{\nu+k}\kappa^2\right)\left(\prod_{\kappa=\nu+k+1}^{2\nu+k+1}\kappa^2\right)z^{-\nu-k-1}. \tag{6}$$

If we set $y=\nu+k+1$ here, we find that

$$f_2(z;\nu) = \sum_{y=\nu+1}^{\infty} R_\nu(y)z^{-y}, \tag{7}$$

where

$$R_\nu(y) = \frac{(y-1)^2\cdots(y-\nu)^2}{y^2(y+1)^2\cdots(y+\nu)^2}. \tag{8}$$

Hence,

$$f_2(z;\nu) = \sum_{y=1}^{\infty} R_\nu(y)z^{-y}. \tag{9}$$

Finally, let

$$\begin{aligned} f_3(z;\nu) &= C_{4,4}^{4,2}\left(z\,\middle|\,\begin{matrix}-\nu, & -\nu, & \nu+1, & \nu+1\\ 0, & 0, & 0, & 0\end{matrix}\right) \\ &= \frac{1}{2\pi i}\int_{L_2}\frac{\Gamma^4(-s)\Gamma^2(\nu+1+s)}{\Gamma^2(\nu+1-s)}z^s\,ds. \end{aligned} \tag{10}$$

If we set

$$H(s) = \Gamma^4(-s)\Gamma^2(\nu+1+s)\Gamma^{-2}(\nu+1-s)z^s \tag{11}$$

we find that

$$f_3(z;\nu) = \sum_{k=0}^{\infty}\operatorname*{res}_{s=-\nu-1-k} H(s). \tag{12}$$

Furthermore,

$$\operatorname*{res}_{s=-\nu-1-k} H(s) = \left(\log z R_\nu(\nu+1+k) - \left(\frac{d}{dy}R_\nu\right)(\nu+1+k)\right)z^{-\nu-1-k}. \tag{13}$$

Hence,

$$f_3(z;\nu) = \log z f_2(z;\nu) + \sum_{y=1}^{\infty} z^{-y} \left(\frac{d}{dy} R_\nu\right)(y). \tag{14}$$

Expanding $R_\nu(y)$ into partial fractions, we obtain

$$R_\nu(y) = \sum_{k=0}^{\nu} \left(\frac{\alpha_{\nu,k}}{(y+k)^2} + \frac{\beta_{\nu,k}}{y+k}\right), \tag{15}$$

where

$$\begin{gathered} \alpha_{\nu,k} = \binom{\nu}{k}^2 \binom{\nu+k}{k}^2, \\ \beta_{\nu,k} = \operatorname*{res}_{y=-k} R_\nu(y) = 2\alpha_{\nu,k} \left(-\sum_{\kappa=k+1}^{\nu+k} \frac{1}{\kappa} + \sum_{\kappa=1}^{k} \frac{1}{\kappa} - \sum_{\kappa=1}^{\nu-k} \frac{1}{\kappa}\right). \end{gathered} \tag{16}$$

From (9) and (15) it follows that

$$\begin{aligned} f_2(z;\nu) &= \sum_{y=1}^{\infty} \left(\sum_{k=0}^{\nu} (\alpha_{\nu,k}(y+k)^{-2} + \beta_{\nu,k}(y+k)^{-1})\right) z^{-y} \\ &= \sum_{k=0}^{\nu} \left(\alpha_{\nu,k} \sum_{y=1}^{\infty} z^{-y}(y+k)^{-2} + \beta_{\nu,k} \sum_{y=1}^{\infty} z^{-y}(y+k)^{-1}\right) \\ &= \sum_{k=0}^{\nu} z^k \left(\alpha_{\nu,k}\left(L_2(z^{-1}) - \sum_{1\le l\le k} z^{-l} l^{-2}\right)\right. \\ &\qquad \left. + \beta_{\nu,k}\left(-\log\left(1-\frac{1}{z}\right) - \sum_{1\le l\le k} z^{-l} l^{-1}\right)\right) \\ &= \alpha(z;\nu) L_2(z^{-1}) + \beta(z;\nu)\left(-\log\left(1-\frac{1}{z}\right)\right) - \varphi(z;\nu), \end{aligned} \tag{17}$$

where

$$L_2(z) = \sum_{l=1}^{\infty} z^l l^{-2}, \tag{18}$$

$$\alpha(z;\nu) = \sum_{0\le k\le \nu} \alpha_{\nu,k} z^k = f_1(z;\nu), \tag{19}$$

$$\beta(z;\nu) = \sum_{0\le k\le \nu} \beta_{\nu,k} z^k, \tag{20}$$

$$\varphi(z;\nu) = \sum_{0\le k\le \nu} \sum_{1\le l\le k} z^{k-l}(\alpha_{\nu,k} l^{-2} + \beta_{\nu,k} l^{-1}). \tag{21}$$

It similarly follows from (14) and (15) that

$$f_3(z;\nu) = (\log z) f_2(z;\nu) + 2\alpha(z;\nu) L_3(z^{-1}) + \beta(z;\nu) L_2(z^{-1}) - \psi(z;\nu), \tag{22}$$

where

$$L_3(z) = \sum_{l=1}^{\infty} z^l l^{-3}, \tag{23}$$

$$\psi(z;\nu) = \sum_{0\le k\le\nu} \sum_{1\le l\le k} z^{k-1}(2\alpha_{\nu,k}l^{-3} + \beta_{\nu,k}l^{-2}). \tag{24}$$

If $\nu \in \mathbf{N}+1$, we let D_ν denote the smallest number in $\mathbf{N}$ with the property that the following inequality holds for every $k = 1,\dots,2\nu$ and for every prime p, $2 \le p \le \nu$:

$$v_p(k^{-1}D_\nu) \ge 0,$$

where v_p is the additive p-adic valuation on the field $\mathbf{Q}$. It is clear that for any $\varepsilon > 0$

$$D_\nu = \prod_{p\le\nu} p^{[\log 2\nu/\log p]} \le e^{(\log 2\nu)\left(\frac{\nu}{\log\nu}+O\left(\frac{\nu}{(\log\nu)^2}\right)\right)} = O(e^{\nu(1+\varepsilon)}). \tag{25}$$

Further, let

$$\begin{aligned} K(z;\nu) &= D_\nu^3\alpha(z;\nu), & L(z;\nu) &= D_\nu^3\beta(z;\nu),\\ M(z;\nu) &= D_\nu^3\varphi(z;\nu), & N(z;\nu) &= D_\nu^3\psi(z;\nu). \end{aligned} \tag{26}$$

Clearly,

$$K(z;\nu) \in \mathbf{Z}[z].$$

We now show that

$$L(z;\nu) \in \mathbf{Z}[z], \qquad M(z;\nu) \in \mathbf{Z}[z], \qquad N(z;\nu) \in \mathbf{Z}[z]. \tag{27}$$

It is sufficient to establish that for any prime $p \in \nu+\mathbf{N}$ we have the inequality $v_p(\beta_{\nu,k}) \ge 0$, $0 \le k \le \nu$. If $\nu + k < p$, this inequality obviously follows from (16). But if $\nu+1 \le p \le \nu+k$, then p^2 is a divisor of the number $\alpha_{\nu,k} = ((\nu+k)!)^2(k!)^{-4}((\nu-k)!)^{-2}$. Thus, since $\kappa \le 2\nu$ and since $p^2 > 2\nu$ if $p \ge \nu+1$, we have $v_p(p^2/\kappa) \ge 0$. Thus, we have proved (27).

Let δ be the differential operator $\delta = z(d/dz)$. From well-known properties of the Meijer functions it follows that

$$(\delta-\nu-1)^2 f_r(z;\nu+1) = (\delta+\nu+1)^2 f_r(z;\nu), \qquad r = 1,2,3, \tag{28}$$

$$(z(\delta+\nu+1)^2(\delta-\nu)^2 - \delta^4)f_r(z;\nu) = 0. \tag{29}$$

Let

$$X_r(z;\nu) = \begin{bmatrix} f_r(z;\nu)\\ \delta f_r(z;\nu)\\ \delta^2 f_r(z;\nu)\\ \delta^3 f_r(z;\nu)\end{bmatrix}, \tag{30}$$

$$B(z;\nu) = \begin{bmatrix} 0 & 1 & 0 & 0 \\ 0 & 0 & 1 & 0 \\ 0 & 0 & 0 & 1 \\ -\frac{\nu^2(\nu+1)^2 z}{z-1} & \frac{2\nu(\nu+1)z}{z-1} & \frac{(2\nu(\nu+1)-1)z}{z-1} & -\frac{2z}{z-1} \end{bmatrix}. \tag{31}$$

From (28)–(31) it follows that

$$(\delta - \nu - 1)^2 X_r(z;\nu+1) = (\delta + \nu + 1)^2 X_r(z;\nu), \tag{32}$$
$$\delta X_r(z;\nu) = B(z;\nu) X_r(z;\nu), \tag{33}$$
$$(\delta - \nu - 1)^2 X_r(z;\nu+1)$$
$$= (\delta B(z;\nu+1) + (B(z;\nu+1) - (\nu+1)E)^2) X_r(z;\nu+1), \tag{34}$$
$$(\delta + \nu + 1)^2 X_r(z;\nu) = (\delta B(z;\nu) + (B(z;\nu) + (\nu+1)E)^2) X_r(z;\nu). \tag{35}$$

Let

$$A(z;\nu) = (\delta B(z;\nu+1) + (B(z;\nu+1) - (\nu+1)E)^2)^{-1}$$
$$\times ((\delta B(z;\nu) + (B(z;\nu) + (\nu+1)E)^2). \tag{36}$$

From (32) and (34)–(36) it follows that

$$X_r(z;\nu+1) = A(z;\nu) X_r(z;\nu), \qquad r = 1,2,3. \tag{37}$$

Further, let

$$T_\nu = \begin{bmatrix} 1 & 0 & 0 & 0 \\ 0 & \nu & 0 & 0 \\ 0 & 0 & \nu^2 & 0 \\ 0 & 0 & 0 & \nu^3 \end{bmatrix}, \tag{38}$$
$$Y_r(z;\nu) = T_\nu^{-1} X_r(z;\nu), \qquad r = 1,2,3, \quad \nu \in \mathbf{N}, \tag{39}$$
$$C(z;\nu) = T_{\nu+1}^{-1} A(z;\nu) T_\nu. \tag{40}$$

Then from (37)–(40) it follows that

$$Y_r(z;\nu+1) = C(z;\nu) Y_r(z;\nu). \tag{41}$$

For every $i = 1,2,3,4$ and $k = 1,2,3,4$ we let $c_{i;k}(z;\nu)$ denote the entry in the ith row and kth column of the matrix $C(z;\nu)$. Performing the appropriate computations in (36)–(40), we obtain the following representations for $c_{i;k}(z;\nu)$,

$i = 1, 2, 3, 4,\ k = 1, 2, 3, 4$:

$$-(\nu+1)^3 c_{1;1}(z;\nu) = (-16z-1)\nu^3 + (-36z-3)\nu^2 + (-24z-3)\nu + (-4z-1), \tag{42}$$
$$-(\nu+1)^3 c_{1;2}(z;\nu) = (-20z-4)\nu^3 + (-16z-8)\nu^2 + (4z-4)\nu, \tag{43}$$
$$-(\nu+1)^3 c_{1;3}(z;\nu) = (8z-8)\nu^3 + (20z-8)\nu^2, \tag{44}$$
$$-(\nu+1)^3 c_{1;4}(z;\nu) = (12z-12)\nu^3, \tag{45}$$
$$-(\nu+1)^3 c_{2;1}(z;\nu) = -z(12\nu^3 + 28\nu^2 + 20\nu + 4), \tag{46}$$
$$-(\nu+1)^3 c_{2;2}(z;\nu) = (-16z-1)\nu^3 + (-16z-2)\nu^2 - \nu, \tag{47}$$
$$-(\nu+1)^3 c_{2;3}(z;\nu) = (4z-4)\nu^3 + (12z-4)\nu^2, \tag{48}$$
$$-(\nu+1)^3 c_{2;4}(z;\nu) = (8z-8)\nu^3, \tag{49}$$
$$-(\nu+1)^3 c_{3;1}(z;\nu) = 4z(-2\nu^3 - 5\nu^2 - 4\nu - 1), \tag{50}$$
$$-(\nu+1)^3 c_{3;2}(z;\nu) = -4z(3\nu^3 + 4\nu^2 + \nu), \tag{51}$$
$$-(\nu+1)^3 c_{3;3}(z;\nu) = -\nu^3 + (4z-1)\nu^2, \tag{52}$$
$$-(\nu+1)^3 c_{3;4}(z;\nu) = 4(z-1)\nu^3, \tag{53}$$
$$-(\nu+1)^3 c_{4;1}(z;\nu) = -4z(\nu^3 + 3\nu^2 + 3\nu + 1), \tag{54}$$
$$-(\nu+1)^3 c_{4;2}(z;\nu) = -8z(\nu^3 + 2\nu^2 + \nu), \tag{55}$$
$$-(\nu+1)^3 c_{4;3}(z;\nu) = -4z(\nu^3 + \nu^2), \tag{56}$$
$$-(\nu+1)^3 c_{4;4}(z;\nu) = -\nu^3. \tag{57}$$

If we successively apply the operators $\delta, \delta^2, \delta^3$ to (17) and (22), and if we take into account that

$$\delta(L_3(z^{-1})) = -L_2(z^{-1}), \qquad \delta(L_2(z^{-1})) = \log(1-1/z),$$
$$\delta\left(\log\left(1-\frac{1}{z}\right)\right) = \frac{1}{z-1}, \qquad \delta\left(\frac{1}{z-1}\right) = \frac{-z}{(z-1)^2},$$
$$\delta\left(\frac{-z}{(z-1)^2}\right) = \frac{z}{(z-1)^2} + \frac{2z}{(z-1)^3},$$

then we can conclude that (41) holds not only for $|z| > 1$ but also for $|z| = 1$, $z \neq 1$.

We now set

$$H_r(z) = \sum_{\nu=0}^{\infty} Y_r(-1;\nu) z^\nu, \qquad r = 1, 2, 3. \tag{58}$$

Then $H_r(z)$, $r = 1, 2, 3$, is a solution of the following system of differential equations:

$$\begin{bmatrix} 1+15z & 16z & -16z & -24z \\ 12z & 1+15z & -8z & -16z \\ 8z & 12z & 1-z & -8z \\ 4z & 8z & 4z & 1-z \end{bmatrix} \delta^3 H(z) + z \begin{bmatrix} 33 & 8 & -28 & 0 \\ 28 & 14 & -16 & 0 \\ 20 & 16 & 14 & 0 \\ 12 & 16 & 4 & 0 \end{bmatrix} \delta^2 H(z)$$
$$+ z \begin{bmatrix} 21 & -8 & 0 & 0 \\ 20 & -1 & 0 & 0 \\ 16 & 4 & 0 & 0 \\ 12 & 8 & 0 & 0 \end{bmatrix} \delta H(z) + z \begin{bmatrix} 3 & 0 & 0 & 0 \\ 4 & 0 & 0 & 0 \\ 4 & 0 & 0 & 0 \\ 4 & 0 & 0 & 0 \end{bmatrix} H(z) = 0. \quad (59)$$

We now look at the singular points of the coefficients of the system which arises as a result of multiplying (59) on the left by the matrix

$$\begin{bmatrix} 1+15z & 16z & -16z & -24z \\ 12z & 1+15z & -8z & -16z \\ 8z & 12z & 1-z & -8z \\ 4z & 8z & 4z & 1-z \end{bmatrix}^{-1}.$$

These singular points make up the set of all numbers $z \neq 0$ for which z^{-1} belongs to the set of all eigenvalues of the matrix

$$V = \begin{bmatrix} -15 & -16 & 16 & 24 \\ -12 & -15 & 8 & 16 \\ -8 & -12 & 1 & 8 \\ -4 & -8 & -4 & 1 \end{bmatrix}. \quad (60)$$

Let

$$U = \begin{bmatrix} 0 & 1 & 0 & 0 \\ 0 & 0 & 1 & 0 \\ 0 & 0 & 0 & 1 \\ -1/2 & 0 & 1 & 0 \end{bmatrix}.$$

It is not hard to see that the matrix V in (60) satisfies

$$V = ((U-E)^{-1}(U+E))^2.$$

Since the eigenvalues of V are the squares of the eigenvalues of $(U-E)^{-1}(U+E)$ and $H_r(z)$, $r = 1, 2, 3$, satisfies (59), it follows that the entries in the column $H_r(z)$ can have singularities only at the points $z = \theta^{-2}$, where θ runs through the roots of the polynomial $(\theta+1)^4 + 16\theta^2$. We let $\theta_1, \theta_2 = \bar{\theta}_1$, $\theta_3, \theta_4 = \bar{\theta}_3$ denote the roots of this polynomial. Then $|\theta_1\theta_3| = 1$. One can see that one of the numbers $|\theta_1|$, $|\theta_3|$ is greater than 21. Suppose that $|\theta_1| > 21$. Then [1]

$$|\theta_2| = |\theta_1| > 20.09 > e^3, \qquad |\theta_3| = |\theta_4| < e^{-3}. \quad (61)$$

Let $h_r(z)$ be the first entry in the column $H_r(z)$. Then

$$h_r(z) = \sum_{\nu=0}^{\infty} f_r(-1, \nu) z^\nu. \quad (62)$$

It is easy to see that $f_r(-1,\nu) = O(1)$, $r = 2,3$, as $v \to \infty$. Hence, the radius of convergence ρ_r, $r = 2,3$, of the series (62) is at least 1, and thus $\rho_r \geq |\theta_1|$, $r = 2,3$. Hence,

$$|f_r(-1,\nu)| = O((20.09)^{-\nu}), \qquad r = 2. \tag{63}$$

From (17), (22), (26), (25), (61), and (63) it follows that

$$\begin{aligned} &\lim_{\nu\to\infty}(-\tfrac{1}{2}K(-1,\nu)\varsigma(2) - L(-1,\nu)\log 2 - M(-1,\nu)) = 0,\\ &\lim_{\nu\to\infty}(-\tfrac{3}{2}K(-1,\nu)\varsigma(3) - L(-1,\nu)\tfrac{1}{2}\varsigma(2) - N(-1,\nu)) = 0. \end{aligned} \tag{64}$$

We now show that for any $\kappa \in \mathbf{N}$ the sequences

$$\alpha(-1,\kappa+\nu), \qquad \nu = 1,2,\ldots, \tag{65}$$

$$\beta(-1,\kappa+\nu), \qquad \nu = 1,2,\ldots, \tag{66}$$

$$\varphi(-1,\kappa+\nu), \qquad \nu = 1,2,\ldots, \tag{67}$$

$$\psi(-1,\kappa+\nu), \qquad \nu = 1,2,\ldots \tag{68}$$

form a linearly independent system over $\mathbf{C}$. It is sufficient to show that the system (65)–(68) is linearly independent over $\mathbf{Q}$. From (4) and (19) it follows that

$$\alpha(-1,\nu) \in \mathbf{Z}. \tag{69}$$

Let p be an arbitrary prime greater than 3. From (16) it follows that

$$v_p(\alpha_{2p;k}) \geq 2, \qquad 1 \leq k \leq p-1, \quad p+1 \leq k \leq 2p-1, \tag{70}$$

$$\alpha_{2p;0} = 1, \tag{71}$$

$$v_p(\alpha_{2p;p} - 36) \geq 1, \tag{72}$$

$$v_p(\alpha_{2p;2p} - 36) \geq 1, \tag{73}$$

$$v_p(\beta_{2p;k}) \geq 1, \qquad 1 \leq k \leq p-1, \quad p+1 \leq k \leq 2p-1, \tag{74}$$

$$v_p(\beta_{2p;0} + 6/p) \geq 0, \qquad v_p(\beta_{2p;p} + 60/p) \geq 0, \tag{75}$$

$$v_p(\beta_{2p;2p} - 66/p) \geq 0. \tag{76}$$

From (4), (19), and (70)–(73) it follows that

$$v_p(\alpha(-1,2p) - 1) > 0. \tag{77}$$

From (20) and (74)–(76) it follows that

$$v_p(\beta(-1,2p) - 138/p) \geq 0. \tag{78}$$

From (21) and (70)–(76) it follows that

$$v_p(\varphi(-1,2p) + 93/p^2) \geq -1. \tag{79}$$

From (24) and (70)–(76) it follows that

$$v_p(\psi(-1,2p) + 114/p^3) \geq -2. \tag{80}$$

From (77)–(80) it follows that for any $\kappa \in \mathbf{N}$ the sequences (65)–(68) form a linearly independent system over $\mathbf{Q}$, and hence over $\mathbf{C}$. We may now proceed to the final stage in the proof of Theorem 1.

Let q be a nonzero rational number for which both of the numbers in (1) are rational; then there exists n_0 such that the numbers

$$n_0\left(\varsigma(2)-\frac{3}{q}\varsigma(3)\right),\quad n_0\left(2\log 2-\frac{1}{q}\varsigma(2)\right),\quad n_0/q$$

are integers. As shown above, the linear combination

$$\begin{aligned}\eta(\nu)&=f_2(-1,\nu)-\frac{1}{q}\operatorname{Im}(f_3(-1,\nu))\\&=\left(\varsigma(2)-\frac{3}{q}\varsigma(3)\right)\frac{\alpha(-1,\nu)}{2}+\left(2\log 2-\frac{1}{q}\varsigma(2)\right)\frac{\beta(-1,\nu)}{-2}\\&\qquad-\varphi(-1,\nu)+\frac{1}{q}\psi(-1,\nu)\end{aligned}$$

takes nonzero values infinitely many times as $\nu\to\infty$. Hence,

$$\begin{aligned}2D_\nu n_0\eta(\nu)=\left(\varsigma(2)-\frac{3}{q}\varsigma(3)\right)n_0K(-1,\nu)+\left(2\log 2-\frac{1}{q}\varsigma(2)\right)n_0L(-1,\nu)\\-n_0M(-1,\nu)+\frac{n_0}{q}N(-1,\nu)\end{aligned}$$

takes nonzero values in $\mathbf{Z}$ infinitely many times as $\nu\to\infty$, and this contradicts (64). Thus, Theorem 1 is proved.

Theorem 2 is a direct consequence of Theorem 1.

REMARK ADDED IN PROOF. I have now obtained the following result. Suppose that

$$\begin{gathered}\rho_1=(5+4\sqrt{2}+2\sqrt{2\sqrt{2}-2}+4\sqrt{2\sqrt{2}+2})e^{-3},\qquad\rho_2=\rho_1e^6,\\\gamma=(\log\rho_2)/\log\rho_1,\qquad\varphi_1(x_1,x_2)=x_1\log 4+x_2\varsigma(2),\\\varphi_2(x_1,x_2)=x_1\varsigma(2)+x_23\varsigma(3),\end{gathered}$$

and $\|d\|$ is the distance from d to $\mathbf{Z}$. Then for any $\varepsilon>0$ there exists $c>0$ such that, if x_1, $x_2\in\mathbf{Z}$ and $x_1^2+x_2^2>0$, then

$$\max_{i=1,2}(\|\varphi_i(x_1,x_2)\|)\geq c(\max(|x_1|,|x_2|))^{-\gamma-\varepsilon}.$$

I have also obtained some other quantitative results.

Acknowledgments. It is hard for me to find the words to describe the support shown me by N. I. Fel′dman or to express the extent of my gratitude to him.

V. G. Sprindzhuk looked through an earlier version of my work and offered valuable assistance; I would like to take this opportunity to thank him.

Yu. V. Nesterenko read the first draft of the manuscript and called my attention to some omissions. I would like to thank him, along with M. S. Agranovich and A. N. Andrianov, for their support.

After reading the original manuscript, the referee of this journal made some profound suggestions, including the possibility of using the analytic theory of differential equations and the Z-transform in connection with the goals of this paper. Using these remarks, I was able not only to shorten the proof but also to improve the exposition. I would like to express my appreciation to the referee.

Last, but not least, I would like to conclude by thanking Professor Yu. I. Zhuravlev for the considerable administrative support he has shown me.

REFERENCES

1. Herbert Bristol Dwight, *Tables of integrals and other mathematical data*, 4th ed., Macmillan, 1961.

2. G. V. Chudnovsky, *Transcendental numbers*, Number Theory, Carbondale, 1979, Lecture Notes in Math., vol. 751, Springer-Verlag, 1979, pp. 45–69.

Added in translation. Let $z \in \mathbf{Q}$, $|z| \geq 1$, $x = 1/z$, $b \in \mathbf{N}$, $bz \in \mathbf{Z}$,

$$\begin{aligned}
\eta_k(z) &= (\sqrt[4]{|z|} + \sqrt{\sqrt{|z|} + (-1)^k})^2 \quad (k = 0, 1),\\
\tilde{\eta}_2(z) &= \sqrt{|z|} + \sqrt{|z| + 1} + \sqrt{2(\sqrt{|z|^2 + |z|} + |z|)},\\
\beta_k(z) &= (\log(e^3 b) + \log \tilde{\eta}_{2[k/2]}(z)) / \log((\tilde{\eta}_k(z))^2/(e^3 b)) \quad (k = 0, 1, 2),\\
\alpha_k(z) &= (\log(\tilde{\eta}_0(z)/\tilde{\eta}_1(z)) / \log((\tilde{\eta}_1(z))^2/(e^3 b)))\\
&\qquad \times (1 - (-1)^k) + \beta_k(z) \quad (k = 0, 1, 2),\\
L_{i,s}(x) &= (i/x - (-1)^i) \sum_{n=1}^{\infty} x^n / n^s \quad (i = 0, 1;\ s \in \mathbf{N};\ |x - 1| + s > 1),
\end{aligned}$$

$$L_{1,1}(1) = 0,$$

$$\begin{aligned}
D_k(b) &= \{y | y \in \mathbf{R}, (-1)^{[k/2]} y > (e^{3/2} b^{1/2} - (-1)^k)^4\\
&\qquad \times ((e^{3/2} b^{1/2} + 1)^4 / (e^3 b + 1)^2)^{[k/2]} / (16 e^3 b)\} \quad (k = 0, 1, 2),\\
&x_1 \in \mathbf{R},\ x_2 \in \mathbf{Q}, \quad |x_1| + |x_2| > 0,\\
&\varphi_i = \varphi_i(x_1, x_2, x) = \tilde{\varphi}_i(z, x_1, x_2) = x_1 L_{2-i,i}(x) + x_2 \binom{i}{1} L_{2-i,i+1}(x)\\
&\qquad (i = 1, 2),
\end{aligned}$$

$$\begin{aligned}
\varphi_3 &= x_1,\\
\hat{\alpha}_0(x) &= \alpha_0(z),\\
\tilde{\alpha}_i(x) &= \alpha_1(z) \quad (i = 1, 2),\\
\tilde{\beta}_i(x) &= \beta_1(z) \quad (i = 1, 2),\\
\tilde{\alpha}_0(x) &= \alpha_2(z),\\
\varepsilon &> 0,
\end{aligned}$$

$\|\psi\|$ ($\psi \in \mathbf{R}$) is the distance from ψ to $\mathbf{Z}$.

THEOREM. *With effective positive* $\gamma_0(z,\varepsilon) = \hat{\gamma}_0(x,\varepsilon)$, $\gamma_1(z,\varepsilon) = \tilde{\gamma}_1(x,\varepsilon)$, $\gamma_2(z,\varepsilon) = \tilde{\gamma}_0(x,\varepsilon)$, $\gamma_i^*(z,x_1,x_2,\varepsilon) = \hat{\gamma}_i(x_1,x_2,x,\varepsilon)$ $(i=1,2)$,

$$\max_{i=1,2,3} \|q\varphi_i\| q^{\alpha_0(z)+\varepsilon} \geq \gamma_0(z,\varepsilon) \quad (x_1 = \log z,\ x_2 = 1,\ z \in D_0(b)),$$

$$\max_{i=1,2,3} \|q\varphi_i\| q^{\beta_k(z)+\varepsilon} \geq \gamma_k^*(z,x_1,x_2,\varepsilon) \quad (q \in \mathbf{N},\ x_2 \neq 0;\ k=1,2;\ z \in D_k(b)), \tag{1}$$

$$\max_{i=1,2} \|\tilde{\varphi}_i(z,x_1,x_2)\|(|x_1|+|x_2|)^{\alpha_k(z)+\varepsilon} \geq \gamma_k(z,\varepsilon)$$
$$(x_1 \in \mathbf{Z},\ x_2 \in \mathbf{Z};\ k=1,2;\ z \in D_k(b)). \tag{2}$$

COROLLARY. *For* $z = 1 \in D_0(1)$ *this theorem implies Apéry's theorem on the irrationality of* $\varsigma(3)$; *for* $z = -1 \in D_2(1)$ *it implies Theorems* 1 *and* 2 *above.*

BIBLIOGRAPHY

1. L. A. Gutnik, *On the irrationality of some quantities containing* $\varsigma(3)$, *Acta Arith.* **42** (1983), no. 3, 255–264; English transl. above.

2. ——, *On a measure of the irrationality of dilogarithms at rational points*, Deposited at VINITI No. 4346-84.

3. ——, *On the rank over* **Q** *of some real matrices*, Deposited at VINITI No. 5736-84.

4. ——, *On the smallness of some linear forms*, Deposited at VINITI No. 6279-B87.

Translated by N. KOBLITZ

Amer. Math. Soc. Transl.
(2) Vol. **140**, 1988

Characteristic Roots of Stochastic Matrices

N. A. DMITRIEV AND E. B. DYNKIN

Abstract. The authors study the location of characteristic roots of stochastic matrices of order n in the complex plane. The results can be interpreted as theorems on the relative position of characteristic roots of an arbitrary matrix with nonnegative elements.

§1. Posing the problem

1. Stochastic matrices appear in probability theory, more precisely, in the theory of Markov chains. Suppose that a certain system can be found in one of the states $\mathscr{E}_1, \dots, \mathscr{E}_n$. Suppose also that from the state $\mathscr{E}_i$ it can pass, in unit time, into the state $\mathscr{E}_j$ with probability a_{ij}. The numbers a_{ij}, defining the Markov chain, are subject to the following conditions:

(1) $a_{ij} \geq 0$ $(i, j = 1, \dots, n)$,

(2) $\sum_{j=1}^{n} a_{ij} = 1$ $(i = 1, \dots, n)$.

A matrix A with elements a_{ij} subject to these two conditions is called *stochastic*.

From the theory of Markov chains stems the problem, posed by A. N. Kolmogorov, of defining the set M_n of characteristic roots of all stochastic matrices of order n. The purpose of this paper is to study this problem.

2. The problem admits a geometric formulation. The complex number λ belongs to M_n if and only if for some $a_{ij} \geq 0$ satisfying the condition $\sum_{j=1}^{n} a_{ij} = 1$ $(i = 1, \dots, n)$, and for some complex numbers x_i (some of which are nonzero), the following identity holds:

$$\lambda x_i = \sum_{j=1}^{n} a_{ij} x_j \qquad (i = 1, \dots, n).$$

These equations express conditions for the polygon coinciding with the convex hull of the points $x_1, \dots, x_n$ to be mapped into itself through multiplication

1980 *Mathematics Subject Classification* (1985 *Revision*). Primary 15A18, 15A51; Secondary 15A48.

Translation of Izv. Akad. Nauk SSSR Ser. Mat. **10** (1946), 167–184; MR **8**, 129.

by λ.(1) So λ belongs to M_n if and only if there exists a convex k-gon Q ($k \leq n$), consisting of more than one point, which is mapped into itself through multiplication by λ. This is true if and only if $|\lambda| \leq 1$. It follows that M_n is a subset of the unit circle centered at the origin. It also follows that $M_n \subset M_{n+1}$.

3. The convex k-gon Q is called *cyclic, generated by* μ, if Q is the convex hull of the points $1, \mu, \mu^2, \ldots$. For example, consider the regular n-gon with vertices $e^{i2\pi k/n}$ ($k = 0, 1, \ldots, n-1$); it is generated by $e^{i2\pi p/n}$ for all p relatively prime to n.

It is clear that the product of two points in Q is in Q; so if $\lambda \in Q$ then $\lambda Q \subset Q$. M_n contains all cyclic k-gons, $k \leq n$. One may suppose that M_n does not contain any other points. All our results confirm the above hypothesis. We prove it completely for $n \leq 5$.

§2. Basic theorem

4. If $\lambda Q \subset Q$ and $0 < \alpha < 1$, then $\alpha\lambda Q \subset \lambda Q \subset Q$; therefore if $\lambda \in M_n$ then $\alpha\lambda \in M_n$. Moreover, it is easy to see that M_n is a closed set. So in order to define M_n, it suffices to find the numbers λ from M_n for which $\beta\lambda \notin M_n$ when $\beta > 1$. We shall call these λ *extremal.*

In §2 we prove the necessary condition for λ to be extremal. We shall give three formulations of this condition: geometric, algebraic, and probabilistic.

5. BASIC THEOREM. *Let the extremal number λ belong to the set $M_n - M_{n-1}$, and let $2\pi p/n \leq \arg\lambda \leq 2\pi(p+1)/n$.*

5.1. *If the convex n-gon Q contains λQ and if $x_1, \ldots, x_n$ are the vertices of Q in counterclockwise order, then*

$$\lambda x_k = \alpha_k x_{k+p} + \beta_k x_{k+p+1} \quad (k = 1, 2, \ldots, n), \quad \alpha_k \geq 0, \quad \beta_k \geq 0, \quad \alpha_k + \beta_k = 1$$

(*the subscripts are modulo n*).(2)

5.2. *If the stochastic matrix A of order n has λ as a characteristic root then it can be put in the form shown schematically in Figure* A *where all elements, except possibly those belonging to the pth and $(p+1)$th diagonals, are zero.*

5.3. *If a system can be found in one of the states $\mathscr{E}_1, \ldots, \mathscr{E}_n$, and if the probabilities a_{ij} of transition from $\mathscr{E}_i$ to $\mathscr{E}_j$ constitute the matrix A, having λ as a characteristic root, then in this listing of states $\mathscr{E}_1, \ldots, \mathscr{E}_n$ the system may pass in unit time from state $\mathscr{E}_i$ only to the states $\mathscr{E}_{i+p}$ or $\mathscr{E}_{i+p+1}$.*

6. The basic theorem will be proved in the geometric form of 5.1. Let $\lambda \in M_n - M_{n-1}$ be a fixed extremal number, and consider only the convex n-gons which are mapped into themselves through multiplication by λ (this will not

(1)Here multiplication of a set Q by λ means that we rotate Q by $\arg\lambda$ and contract it with coefficient $|\lambda|$.

(2)α_i and β_j cannot be zero simultaneously; this condition implies stronger versions of 5.2 and 5.3, but these will not be used.

be repeated every time). To each such n-gon λQ we assign four integers $\alpha(Q)$, $\beta(Q)$, $\beta_1(Q)$, and $\beta_2(Q)$, defined as follows:

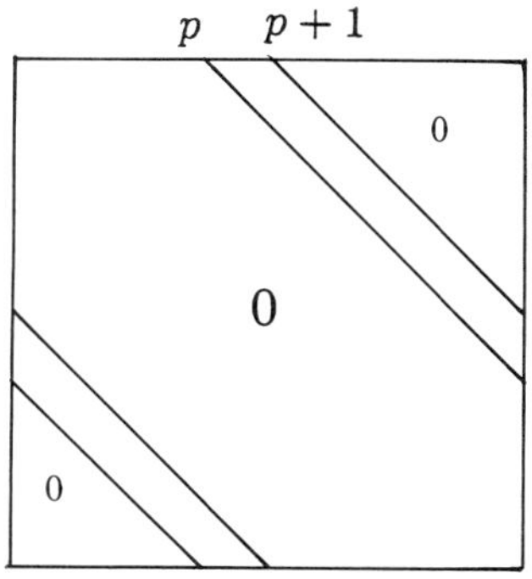

FIGURE A
STOCHASTIC MATRIX WITH AN "EXTREMAL" ROOT

(1) $\alpha(Q)$ is the number of vertices of the n-gon λQ lying in the interior of Q.

(2) The number of sides of the n-gon Q disjoint from λQ is denoted by $\beta(Q)$, $\beta_1(Q)$, or $\beta_2(Q)$, depending on whether we include in the side both endpoints, just the first endpoint, or just the second endpoint, respectively. That is, whether we are dealing with sides of the type $[x\ y]$, $[x\ y)$, or $(x\ y]$.(3)

We have three theorems. For each Q:

THEOREM I. $\alpha(Q) = 0$.

THEOREM II. $\beta(Q) = 0$.

THEOREM III. *Either* $\beta_1(Q) = 0$ *or* $\beta_2(Q) = 0$.

All three proofs are identical. The last one is basically equivalent to Theorem 5.1. Some definitions and lemmas precede the proofs of Theorems I, II, and III. We will denote by the same letter the elements (vertices, sides, etc.) which correspond to each other in the polygons λQ and Q. We shall put a tilde above the letters denoting elements of λQ. We denote the contour of the n-gon Q by $K(Q)$. Segments of $K(Q)$ or $K(\lambda Q)$ which do not meet each other are called *free*.

7. LEMMA I. *Basic Theorem* 5.1 *and therefore Theorems* I, II, *and* III *are satisfied for the polygon* Q *if*:

(3)The endpoints are in counterclockwise order; that is, the first letter, x, of the interval symbolizes the first endpoint.

7.1. *for some sequence of its vertices,* $x_0, x_1, \ldots, x_{p-1}$, *we have* $\lambda x_0 = x_1$, $\lambda x_1 = x_2, \ldots, \lambda x_{p-1} = x_0$ *and*

7.2. *for some sequence of its sides,* $a_0, a_1, \ldots, a_{p-1}$, *we have* $\lambda a_0 \subset a_1, \lambda a_1 \subset a_2, \ldots, \lambda a_{p-1} \subset a_0$.

PROOF. In the case 7.1, $\lambda^p x_0 = x_0$. No polygon is mapped into itself by compression and by rotation around its vertices. Therefore $x_0 \neq 0$, and $\lambda^p = 1$.

In the case 7.2, $\lambda^p a_i \subset a_i$ for $i = 0, 1, \ldots, p-1$; so either $\lambda^p = 1$ or $0 \in a_i$ for $i = 0, 1, \ldots, p-1$. This last condition is possible only if $a_0 = a_1 = \cdots = a_{p-1}$. But then $\lambda a_0 \subset a_0$ and $\lambda \in M_2$. It is obvious that $M_2 = [-1, +1]$; since λ is extremal, $\lambda = 1$ or $\lambda = -1$, and so $\lambda^2 = 1$.

Therefore in all cases $\lambda^k = 1$ for some k. Since $|\lambda| = 1$, the area of λQ equals the area of Q (and $\lambda Q \subset Q$); therefore $\lambda Q = Q$. If y is a vertex of Q, so are the points $\lambda y, \lambda^2 y, \ldots$. The convex hull of the points $y, \lambda y, \lambda^2 y, \ldots$ is a regular polygon which is mapped onto itself through multiplication by λ, and for which the number of vertices is not more than n. When it is equal to n, for $\lambda \notin M_{n-1}$, the polygon coincides with Q. If $\lambda = e^{i2\pi p/n}$ and $y_1, \ldots, y_n$ are vertices of Q, then $\lambda y_k = y_{k+p}$. So Theorem 5.1 holds in both cases.

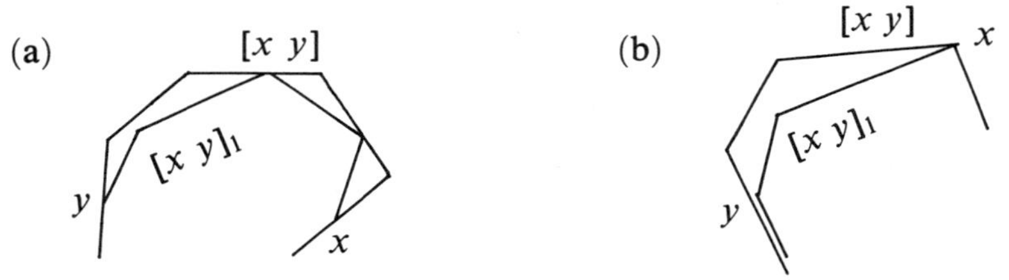

FIGURE 1

8. LEMMA II. *Let the points* x *and* y *belong to the intersection of* $K(Q)$ *and* $K(\lambda Q)$. *Let* $[x\ y]$ *and* $[x\ y]_1$ *be the segments*(4) *defined by these points on* $K(Q)$ *and* $K(\lambda Q)$, *respectively (see Figure* 1(a)). *If* p *equals the number of vertices of* Q *belonging to* $[x\ y]$ *and* q *equals the number of vertices of* λQ *belonging to* $[x\ y]_1$, *then* $p \leq q$.

$p = q$ *in all cases when each of the points* x *and* y *is either a vertex of both* Q *and* λQ *or of neither (see Figure* 1(b)).

PROOF. We remove from the polygon the region bounded by $[x\ y]$ and $[x\ y]_1$ (Figure 1). We thus obtain a polygon Q' satisfying the condition $\lambda Q \subset Q' \subset Q$, whence $\lambda Q' \subset \lambda Q \subset Q'$. We recall that the number of vertices of Q' is $n - p + q$. Since $\lambda \notin M_{n-1}$ we have $n - p + q \geq n$; that is, $p \leq q$.

To prove the second part of the lemma, we construct a polygon $\lambda Q''$ by adding to λQ the region bounded by $[x\ y]$ and $[x\ y]_1$ (Figure 1(b)). Then $\lambda Q \subset \lambda Q'' \subset Q$,

(4) *Editor's note.* $[x\ y]$, which is called a segment in the text, is in fact a polygonal line.

and thus $\lambda Q'' \subset Q \subset Q''$. The number of vertices of $\lambda Q''$ (and hence of Q'') in the cases mentioned in the second part of the lemma is equal to $n - q + p$.

Since $\lambda \notin M_{n-1}$, we have $n - q + p \geq n$; that is, $p \geq q$. Therefore $p = q$.

9. LEMMA III. *Let x and y be vertices of Q and let $(x\ y)$ consist of not less than two sides of Q. Then $(x\ y)$ and the corresponding segment $(\tilde{x}\ \tilde{y})$ of $K(\lambda Q)$ are either both free or both not free.*

LEMMA IV. *Let x, y, and z be three consecutive vertices of Q. If $[y\ z]$ is free and $\tilde{y} \in K(Q)$, then $[\tilde{x}\ \tilde{y}] \subset K(Q)$.*

We shall prove Lemmas III and IV first with the weakened assumption that for an arbitrary n-gon Q', $\alpha(Q') \leq \alpha(Q)$. These weakened lemmas will prove to be sufficient for the proof of Theorem I, from which it follows that our assumption does not impose any further restrictions on Q.

9.1. PROOF OF LEMMA III. Let $(x\ y)$ be free. We construct a new n-gon Q' by replacing in Q the segment $(x\ y)$ by the segment $(x\ y)'$, which has the same number of sides and the same endpoints as $(x\ y)$, and lies entirely in Q but outside λQ (Figure 2).

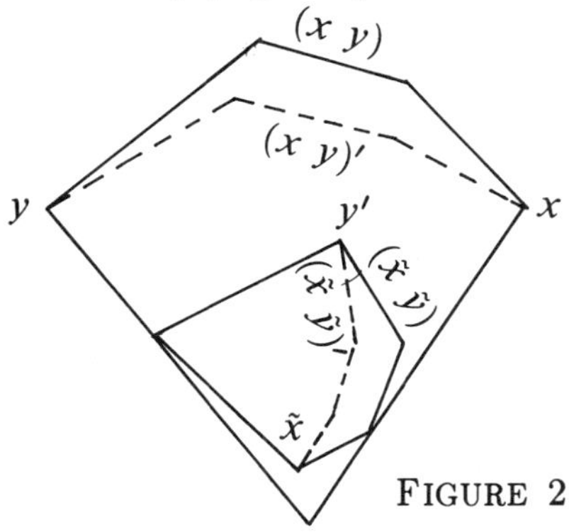

FIGURE 2

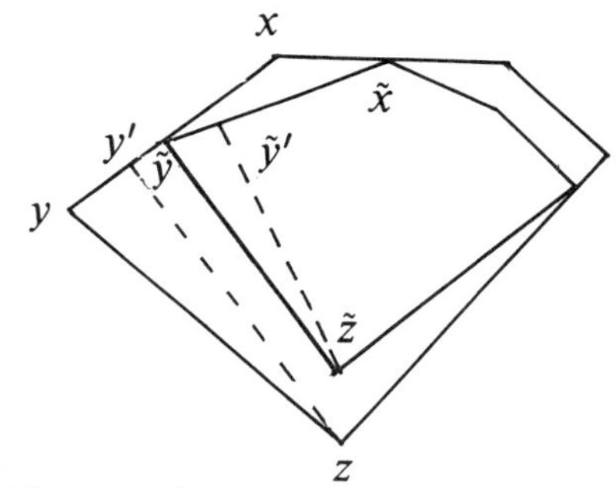

FIGURE 3

Then $\lambda Q'$ is obtained from λQ by replacing $(\tilde{x}\ \tilde{y})$ by $(\tilde{x}\ \tilde{y})' = \lambda(x\ y)'$, lying within λQ. The segment $(\tilde{x}\ \tilde{y})'$ of the contour $K(\lambda Q')$ is free. On the other hand, $\lambda Q' \subset \lambda Q \subset Q'$. Thus $\alpha(Q') \leq \alpha(Q)$, and no vertex of $(\tilde{x}\ \tilde{y})$ can lie on $K(Q)$. The second part of the lemma is proved analogously.

9.2. PROOF OF LEMMA IV. Let Q' be the polygon obtained from Q by removing the triangle yzy', where $y' \in (x\ y)$ (Figure 3). The segment $[y\ z)$ is free; thus we can take y' sufficiently close to y so that the triangle yzy' does not meet λQ; then $Q' \supset \lambda Q$ and $\lambda Q' \subset Q'$. If $[\tilde{x}\ \tilde{y}] \not\subset K(Q)$, then $\tilde{y}' \notin K(Q)$ and therefore $\tilde{y}' \notin K(Q')$. But this contradicts the assumption that $\alpha(Q') \leq \alpha(Q)$.

10. PROOF OF THEOREM I. Suppose that for each n-gon Q', $\alpha(Q') \leq \alpha(Q)$. Then $\alpha(Q) = 0$.

Suppose $\alpha(Q) \neq 0$; let $\tilde{w}$ be a vertex of λQ lying inside Q. The maximal free segment of the contour $K(\lambda Q)$ containing $\tilde{w}$ does not coincide with the entire contour $K(\lambda Q)$; if it did, for sufficiently small $\varepsilon > 0$, we would have

$(1+\varepsilon)\lambda Q \subset Q$, and λ would not be an extremal number. Thus the maximal free segment is of the form $(\tilde{y}_0\ \tilde{z}_0)$, where $\tilde{y}_0$ and $\tilde{z}_0$ are vertices of the n-gon λQ lying on $K(Q)$. By Lemma III, the segment $(y_0\ z_0)$ of $K(Q)$ is free but the segments $(x_0\ z_0)$ and $(y_0\ u_0)$ are not free (x_0 being the vertex of Q immediately preceding y_0, and u_0 being the vertex of Q immediately following z_0). It follows that some vertices of λQ lie on $(x_0\ y_0]$. Denote by $\tilde{y}_1$ the one lying nearest y_0. Similarly, some vertices of λQ lie on $[z_0\ u_0)$. Denote by $\tilde{z}_1$ the one lying nearest z_0. The segment $(\tilde{y}_1\ \tilde{z}_1)$ of $K(\lambda Q)$ is free; the number of sides [of the n-gon] of $(\tilde{y}_1\ \tilde{z}_1)$ is not less than the number of sides of $(y_0\ z_0)$ (by Lemma II) and a fortiori not less than two. The segment $(\tilde{y}_1\ \tilde{z}_1)$ has thus all the properties of the initial segment $(\tilde{y}_0\ \tilde{z}_0)$; therefore we can construct a segment $(\tilde{y}_2\ \tilde{z}_2)$ from $(\tilde{y}_1\ \tilde{z}_1)$ in the same way as we constructed $(\tilde{y}_1\ \tilde{z}_1)$ from $(\tilde{y}_0\ \tilde{z}_0)$, and so on.

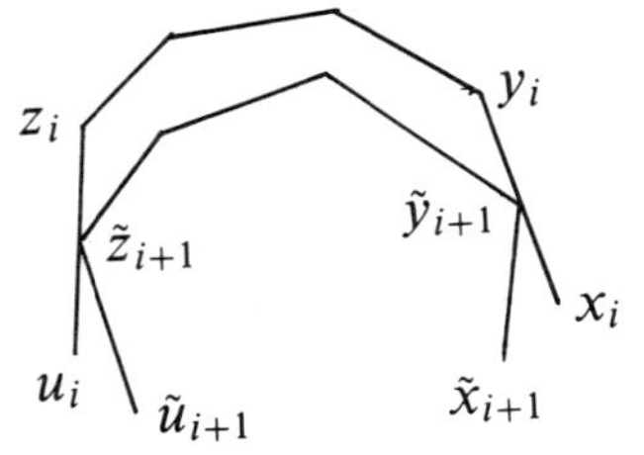

FIGURE 4

Repeating this construction, we define the free segments $(\tilde{y}_3\ \tilde{z}_3), \dots, (\tilde{y}_i\ \tilde{z}_i), \dots$ and at some step p we get the segment $(\tilde{y}_p\ \tilde{z}_p)$ equal to $(\tilde{y}_0\ \tilde{z}_0)$ (see Figure 4). Therefore $\tilde{y}_{i+1} \in (x_i\ y_i]$ $(i = 0, 1, \dots, p-1)$, where x_i is the vertex of Q immediately preceding y_i. We also recall that, since the number of sides of $(y_{i+1}\ z_{i+1})$ is not less than the number of sides of $(y_i\ z_i)$, all the segments $(y_0\ z_0)$, $(y_1\ z_1), \dots, (y_{p-1}\ z_{p-1})$ consist of an equal number of sides.

Four types of positions in the plane are possible for the sides $a_i = (x_i\ y_i]$ and $\tilde{a}_{i+1} = (\tilde{x}_{i+1}\ \tilde{y}_{i+1}]$ relative to one another:(5)

(1) $y_i = \tilde{y}_{i+1}$, $a_i \not\supset \tilde{a}_{i+1}$ (2) $y_i = \tilde{y}_{i+1}$, $a_i \supset \tilde{a}_{i+1}$ (3) $y_i \neq y_{i+1}$, $a_i \supset \tilde{a}_{i+1}$ (4) $y_i \neq \tilde{y}_{i+1}$, $a_i \not\supset \tilde{a}_{i+1}$

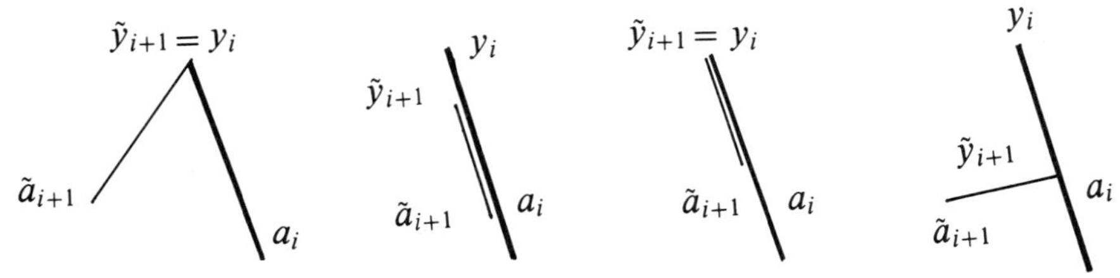

We associate subscript i with one of the classes K_1, K_2, K_3, or K_4, depending on which of the four types of position a_i and $\tilde{a}_{i+1}$ have [relative to one another].

(5) *Editor's note.* A thin line next to a thicker line denotes overlapping sides.

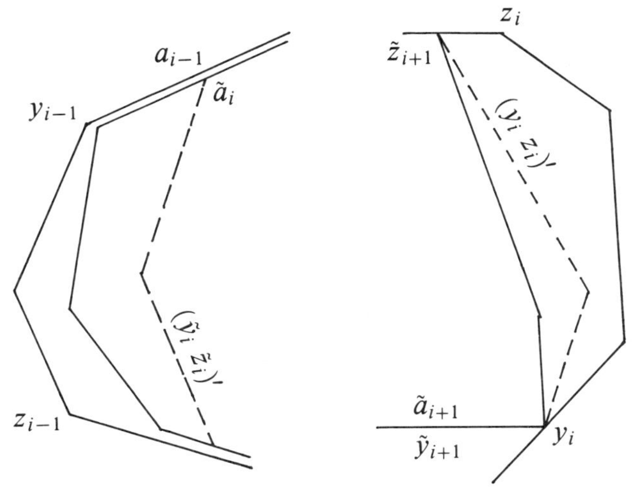

FIGURE 5

We will show that by somewhat changing the n-gon Q, it is always possible to have class K_4 empty. Let $i \in K_4$. We construct an n-gon Q' (see Figure 5), by replacing in $K(Q)$ the segment $(y_i\ z_i)$ by $(y_i\ z_i)'$ near the segment $(\tilde{y}_{i+1}\ \tilde{z}_{i+1})$ of $K(\lambda Q)$, where $(y_i\ z_i)'$ has the same number of sides and the same endpoints as $(\tilde{y}_{i+1}\ \tilde{z}_{i+1})$, and lies outside λQ but inside Q. Then $\lambda Q' \subset Q'$ and $\alpha(Q') = \alpha(Q)$. By passing from Q to Q', the number i falls into the class K_1 and the number $i-1$ into K_3; the other subscripts do not change class. Repeating this construction several times if necessary, we obtain a polygon Q_1 for which $\alpha(Q_1) = \alpha(Q)$ and class K_4 is empty.

For the polygon Q_1, one of the classes K_1 or K_3 is empty. Suppose not. By Lemma IV, no subscript from K_3 can immediately follow a subscript from K_1. Thus they must be separated by some subscripts from K_2. Now we will exhibit a transformation of the n-gon Q_1 which decreases the number of the intermediate subscripts from K_2. By repeating such a transformation we can reduce the number of K_2-subscripts to zero, thus contradicting Lemma IV. So in order to show that one of K_1 or K_3 is empty it remains only to define the transformation of Q_1 mentioned above.

Let $i \in K_1$, $i+1, \ldots, k-1 \in K_2$, and $k \in K_3$. In Q_1 we replace the vertex y_k by $y_k' \in (\tilde{y}_{k+1}\ y_k)$, sufficiently close to y_k so that the resulting n-gon Q_1' contains λQ (see Figure 6). Then $Q_1' \supset \lambda Q_1'$ and for this n-gon, $k-1 \in K_3$, $k \in K_3$; the other subscripts stay in the same classes.

If K_1 is empty for Q_1, then $\tilde{a}_{i+1} = \lambda a_{i+1} \subset a_i$ $(i = 0, 1, \ldots, p-1)$. If K_3 is empty, then $\tilde{y}_{i+1} = \lambda y_{i+1} = y_i$ $(i = 0, 1, \ldots, p-1)$. In both cases, by Lemma I, we find that Theorem I holds, i.e., $\alpha(Q_1) = 0$. And since $\alpha(Q_1) = \alpha(Q)$, this contradicts the assumption that $\alpha(Q) > 0$.

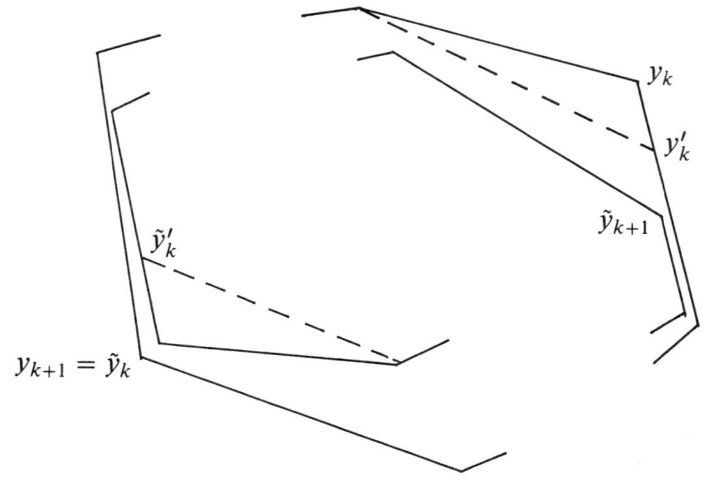

FIGURE 6

11. PROOF OF THEOREM II. Suppose that for each n-gon Q', $\beta(Q') \le \beta(Q)$. Suppose also that $\beta(Q) > 0$. Let $[y_0\ z_0]$ be a free side of the n-gon Q, and let $[x_0\ y_0]$ and $[z_0\ u_0]$ be sides of Q joining $[y_0\ z_0]$. By Theorem I, $\tilde{y}_0$ and $\tilde{z}_0$ lie on $K(Q)$. Hence, by Lemma IV, $[\tilde{x}_0\ \tilde{y}_0] \subset [x_1\ y_1]$ and $[\tilde{z}_0\ \tilde{u}_0] \subset [z_1\ u_1]$, where $[x_1\ y_1]$ and $[z_1\ u_1]$ are sides of Q.

It follows easily from the second part of Lemma II that the vertices y_1 and z_1 are endpoints of a side. We shall show that the side $[y_1\ z_1]$ is free. We remove from Q a sufficiently narrow strip parallel to $[y_0\ z_0]$ (see Figure 7). We thus obtain an n-gon Q' for which $\lambda Q' \subset Q'$ and $[y_1\ z_1]$ is a free side. If $[y_1\ z_1]$ is not a free side of Q_1, then $\beta(Q) < \beta(Q')$, contradicting the hypothesis. We construct the segment $x_2y_2z_2u_2$ of the n-gon Q from the free side $[y_1\ z_1]$ in the same way as we constructed $x_1y_1z_1u_1$ (from $[y_0\ z_0]$).

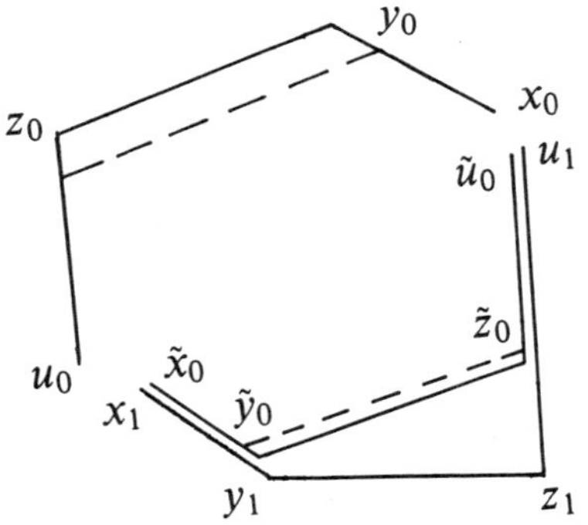

FIGURE 7

Repeating this construction we get a sequence of three-sided segments $x_iy_iz_iu_i$ of Q with $[y_i\ z_i]$ free and $[\tilde{x}_{i-1}\ \tilde{y}_{i-1}] \subset [x_i\ y_i]$, $[\tilde{z}_{i-1}\ \tilde{u}_{i-1}] \subset [z_i\ u_i]$. Using Lemma I, we see that Theorem II holds for Q by contradiction.

12. PROOF OF THEOREM III. Suppose that for each n-gon Q', $\beta_1(Q')\beta_2(Q') \le \beta_1(Q)\beta_2(Q)$. Let $\beta_1(Q)\beta_2(Q) > 0$. We choose a pair of free sides $[y_0\ z_0)$ and $(u_0\ v_0]$ in such a way that the segment $(z_0\ u_0)$ has the smallest possible number of sides. We will show that the vertices $\tilde{z}_0$ and $\tilde{u}_0$ coincide with certain vertices z_1 and u_1 of the n-gon Q, and the sides $[y_1\ z_1)$ and $(u_1\ v_1]$ are free. Applying this result to the pair of free sides $[y_1\ z_1)$ and $(u_1\ v_1]$ (by Lemma II, the segment $(z_1\ u_1)$ has the same number of sides as $(\tilde{z}_0\ \tilde{u}_0)$, namely k), we see that $\tilde{z}_1$ and $\tilde{u}_1$ coincide with vertices z_2 and u_2 of the n-gon Q, and that $[y_2\ z_2)$ and $(u_2\ v_2]$ are free.

Repeating the same construction, we obtain a sequence of vertices z_i of the n-gon Q such that $\tilde{z}_i = z_{i+1}$, and, by Lemma I, we arrive at a contradiction to the assumed inequality $\beta_1(Q)\beta_2(Q) > 0$.

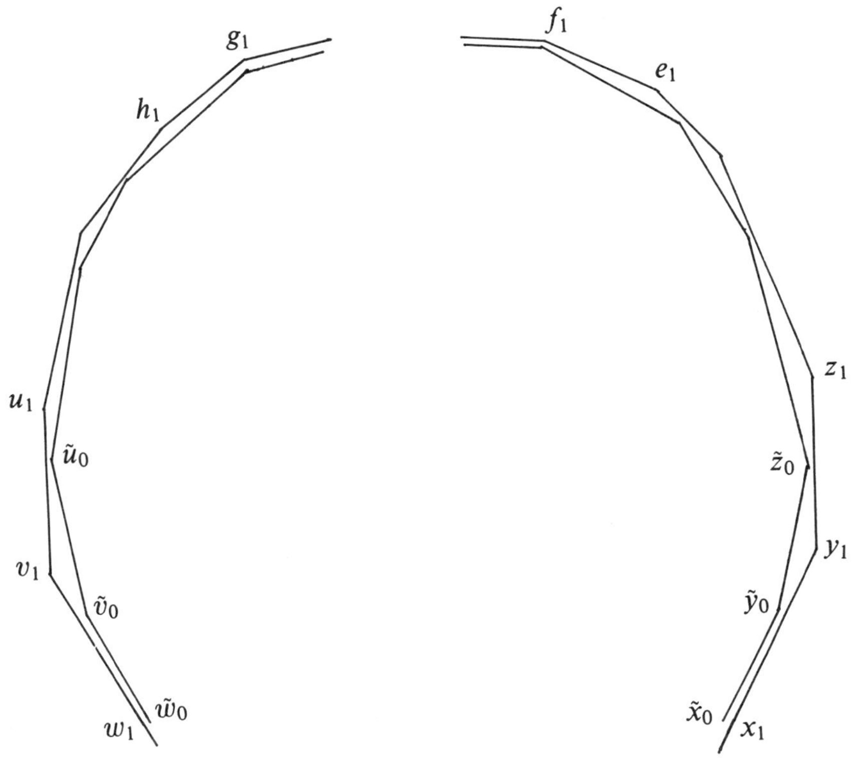

FIGURE 8

Thus it remains to show that $\tilde{z}_0 = z_1$, $\tilde{u}_0 = u_1$, and $[y_1\ z_1)$ and $(u_1\ v_1]$ are free. By Lemma IV, $[\tilde{x}_0\ \tilde{y}_0]$ is contained in some side of the n-gon Q. Call this side $[x_1\ y_1]$. Analogously, call $[v_1\ w_1]$ the side containing $[\tilde{v}_0\ \tilde{w}_0]$ (see Figure 8). If $\tilde{y}_0 \neq y_1$, then it follows immediately from Theorem II that $\tilde{z}_0 \in (y_1\ z_1]$; if $\tilde{y}_0 = y_1$, then by a sufficiently small displacement of the vertex y_0 of the n-gon Q on the side $(x_0\ y_0)$, we obtain an n-gon Q' for which $\lambda Q' \subset Q'$ and $\tilde{y}_0 \neq y_1$. Analogously we can show that $u_0 \in [u_1\ v_1)$.

From Lemma II it follows that the number of sides of $[y_1\ v_1]$ equals $k+2$, the number of sides of $[y_0\ v_0]$.

Suppose that none of the vertices of $[z_1\ u_1]$ lie on $K(\lambda Q)$. Then, inside each side of this segment lies at least one vertex of λQ (by Theorem II). Thus not less than k vertices of λQ lie on $(z_1\ u_1)$. These vertices belong to the segment $[\tilde{z}_0\ \tilde{u}_0]$, which has a total of $k+1$ vertices. Thus all vertices of this segment other than, maybe, one of the endpoints lie on $(z_1\ u_1)$. But this is impossible, since we showed that $\tilde{z}_0 \in (y_1\ z_1]$ and $\tilde{u}_0 \in [u_1\ v_1)$. Consequently, some of the vertices of $[z_1\ u_1]$ lie on the contour $K(\lambda Q)$.

Let f_1 be the first and g_1 the last of these vertices. From Lemma II it follows that $[z_1\ f_1]$ and $[\tilde{z}_0\ f_1]$ have the same number of sides. Since none of the vertices of $[z_1\ f_1]$, excluding f_1, lie on $K(\lambda Q)$, by Theorem II each side of the segment, except maybe the side $(e_1\ f_1)$ preceding f_1, contains at most one vertex of λQ in its interior. Consequently, of the $q+1$ vertices of the segment $[\tilde{z}_0\ f_1]$, at least $q-1$ lie on $(z_1\ e_1)$, one, $\tilde{z}_0$, belongs to $(y_1\ z_1]$, and one coincides with f_1. Thus the side $[e_1\ f_1)$ is free. We can show, in a similar way, that the side $(g_1\ h_1]$, following g_1, is free. By our assumption on the number of sides of $[z_0\ u_0]$, the number of sides of $[z_1\ u_1]$ of the contour $K(Q)$ is at least k. Since $[f_1\ g_1]$ is contained in $[z_1\ u_1]$, which has k sides, we see that $[f_1\ g_1] = [z_1\ u_1]$; that is, z_1 and u_1 lie on $K(\lambda Q)$, and therefore $z_1 = \tilde{z}_0$ and $u_1 = \tilde{u}_0$. The side $[y_1\ z_1)$ is necessarily free, since otherwise, by a sufficiently small displacement of the vertex y_0 of the n-gon Q on $(x_0\ y_0)$, we would obtain an n-gon Q' for which $\lambda Q' \subset Q'$ and $\beta_1(Q')\beta_2(Q') > \beta_1(Q)\beta_2(Q)$, which is impossible. By analogous considerations the side $(u_1\ v_1]$ is also free.

13. Proof of Basic Theorem 5.1. By Theorem III, either $\beta_1(Q) = 0$ or $\beta_2(Q) = 0$. Let $\beta_1(Q) = 0$. Then, if $x_1, \dots, x_n$ are vertices of Q, each of the sides $[x_i\ x_{i+1})$ $(i = 1, \dots, n)$ contain vertices of λQ, for all n. Consequently each of the sides $[x_i\ x_{i+1})$ contains one vertex of λQ and, for some p and for $k = 1, \dots, n, \lambda x_k \in [x_{k+p}\ x_{k+p+1})$. Let $\beta_2(Q) = 0$. We would then arrive at $\lambda x_k \in (x_{k+p}\ x_{k+p+1}]$ $(k = 1, \dots, n)$. We can combine both results in the following statement:

$$\lambda x_k = \alpha_k x_{k+p} + \beta_k x_{k+p+1}, \quad \alpha_k \geq 0, \quad \beta_k \geq 0, \quad \alpha_k + \beta_k = 1 \quad (k = 1, 2, \dots, n).$$

To complete the proof it remains to show that $2\pi p/n \leq \arg\lambda \leq 2\pi(p+1)/n$. Denote by φ_k (see Figure 9) the angle by which we must rotate the side $[x_k\ x_{k+1}]$ to get $[x_{k+1}\ x_{k+2}]$. Through multiplication by λ, the side $[x_k\ x_{k+1}]$ is rotated by $\arg\lambda$ and therefore its endpoints will fall on $[x_{k+p}\ x_{k+p+1}]$ and $[x_{k+p+1}\ x_{k+p+2}]$. Consequently,

$$\varphi_k + \dots + \varphi_{k+p-1} \leq \arg\lambda \leq \varphi_k + \dots + \varphi_{k+p-1} + \varphi_{k+p} \qquad (k = 1, 2, \dots, n).$$

Summing these inequalities over k, and noting that $\sum_1^n \varphi_k = 2\pi$, we obtain

$$2\pi p \leq n \arg\lambda \leq 2\pi(p+1).$$

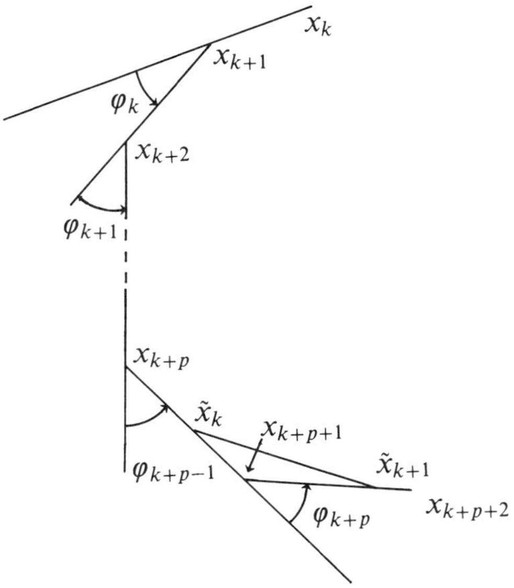

FIGURE 9

§3. Study of the region M_n

14. THEOREM IV. *If λ is an extremal number belonging to the intersection of $M_n - M_{n-1}$ with the angle $\langle 0, 2\pi/n\rangle$*,(6) *then λ lies on the segment joining the points 1 and $e^{2\pi i/n}$. The regular n-gons centered at 0 are the only convex n-gons mapped into themselves through multiplication by λ.*(7)

PROOF. If $\lambda Q \subset Q$ and if $x_1, \ldots, x_n$ denote the vertices of Q in counterclockwise order, then by the Basic Theorem 5.1,

$$\lambda x_k = \alpha_k x_k + \beta_k x_{k+1}, \qquad \alpha_k \geq 0, \quad \beta_k \geq 0, \quad \alpha_k + \beta_k = 1 \quad (k = 1, \ldots, n).$$

We construct the n-gon Q' with vertices $y_1, \ldots, y_n$ by putting

$$y_k = x_k \quad (k \neq i), \qquad y_i = \varepsilon x_{i-1} + (1-\varepsilon) x_i, \quad 0 < \varepsilon < 1.$$

A simple computation shows that

$$\begin{aligned} \lambda y_k &= \alpha_k y_k + \beta_k y_{k+1} \quad (k = 1, \ldots, i-2, i+1, \ldots, n), \\ \lambda y_{i-1} &= \alpha'_{i-1} y_{i-1} + \beta'_{i-1} y_i, \\ \lambda y_i &= \omega'_i y_{i-1} + \alpha'_i y_i + \beta'_i y_{i+1}, \end{aligned}$$

(6) $\langle \varphi_1, \varphi_2 \rangle$ is the set of λ's for which $\varphi_1 \leq \arg \lambda \leq \varphi_2$.

(7) A direct proof of Theorem IV that does not use the idea of extremal number nor Theorem 5 was published in [2].

where

$$\alpha'_{i-1} = \alpha_{i-1} - \frac{\varepsilon}{1-\varepsilon}\beta_{i-1}, \qquad \alpha'_i = \alpha_i + \frac{\varepsilon}{1-\varepsilon}\beta_{i-1} \geq 0,$$
$$\beta'_{i-1} = \frac{1}{1-\varepsilon}\beta_{i-1} \geq 0, \qquad \beta'_i = (1-\varepsilon)\beta_i \geq 0,$$
$$\omega'_i = \varepsilon\left(\alpha_{i-1} - \alpha_i - \frac{\varepsilon}{1-\varepsilon}\beta_{i-1}\right).$$

If the inequalities $\alpha'_{i-1} > 0$ and $\omega'_i > 0$ were satisfied simultaneously, then $Q' \supset \lambda Q'$ and the vertex λy_i of the n-gon $\lambda Q'$ would lie within Q'. This contradicts Theorem I. Let $\alpha_{i-1} > 0$; then, for a sufficiently small ε, $\alpha'_{i-1} > 0$ and therefore $\omega'_i \leq 0$. It follows that the condition $\alpha_{i-1} > 0$ implies $\alpha_{i-1} - \alpha_i \leq 0$; that is, $\alpha_i \geq \alpha_{i-1}$. Thus, either $\alpha_1 = \cdots = \alpha_n = 0$, or, for some q, $\alpha_q > 0$ and $\alpha_q \leq \alpha_{q+1} \leq \cdots \leq \alpha_n \leq \alpha_1 \leq \cdots \leq \alpha_q$; that is, $\alpha_1 = \cdots = \alpha_n = \alpha$ and $\beta_1 = \cdots = \beta_n = \beta$. Thus in all cases:

$$\lambda x_k = \alpha x_k + \beta x_{k+q}, \qquad \alpha, \beta \geq 0, \quad \alpha + \beta = 1 \quad (k = 1, \ldots, n);$$

whence we conclude that

$$\frac{\lambda - \alpha}{\beta} x_k = x_{k+1} \quad (k = 1, 2, \ldots, n),$$
$$\left(\frac{\lambda - \alpha}{\beta}\right)^n x_1 = x_1, \qquad \left(\frac{\lambda - \alpha}{\beta}\right)^n = 1.$$

Consequently, $(\lambda - \alpha)/\beta = e^{2\pi i/n}$ and $\lambda = \alpha + \beta e^{2\pi i/n}$.

Thus $x_{k+1} = x_1 e^{2\pi i k/n}$ $(k = 1, \ldots, n-1)$. The equality $(\lambda - \alpha)/\beta = e^{2\pi i p/n}$ is impossible for $p \neq 1$, since the points $x_1, \ldots, x_n$ are in counterclockwise order.

15. Theorem IV defines M_n completely in the angle $\langle 0, 2\pi/n\rangle$. In order to study M_n in the angle $\langle 2\pi/n, 4\pi/n\rangle$ we revert to the probabilistic interpretation of our problem. If the probability of a system going from state $\mathscr{E}_i$ to state $\mathscr{E}_j$ in one unit of time is a_{ij}, then we denote by $a_{ij}^{(2)}$ the probability that the system goes from state $\mathscr{E}_i$ to state $\mathscr{E}_j$ in two units of time, where

$$a_{ij}^{(2)} = \sum_{k=1}^{n} a_{ik} a_{kj}.$$

If A is the matrix with elements a_{ij}, then its square A^2 is the matrix with elements $a_{ij}^{(2)}$. In general we denote by $a_{ij}^{(m)}$ the probability that the system goes from state $\mathscr{E}_i$ to state $\mathscr{E}_j$ in m units of time; these probabilities form the matrix A^m.

Suppose that the system is initially in any of $\mathscr{E}_1, \ldots, \mathscr{E}_n$ with equal probability. The probability that the system will return to its initial state in one unit of time is denoted by σ_1, where

$$\sigma_1 = \frac{1}{n} \sum_{i=1}^{n} a_{ii} = \frac{1}{n} \operatorname{tr}(A);$$

but $\operatorname{tr}(A) = \sum_1^n \lambda_i$, so

$$\sigma_1 = \frac{1}{n}\sum_{i=1}^n \lambda_i.$$

The characteristic roots of A^m are $\lambda_1^m, \dots, \lambda_n^m$; if the probability that the system will return to its initial state in m units of time is denoted by σ_m, then

$$\sigma_m = \frac{1}{n}\sum_{i=1}^n a_{ii}^{(m)} = \frac{1}{n}\sum_{i=1}^n \lambda_i^m.$$

16. THEOREM V. *Let the extremal number λ belong to the intersection of $M_n - M_{n-1}$ with the angle $\langle 2\pi/n, 4\pi/n\rangle$, and let $n = 2q-1$ be an odd number.*(8) *Then λ generates a cyclic n-gon and is defined by the following equations:*

$$\lambda^n = \alpha + \beta\lambda, \qquad \alpha \geq 0,\ \beta \geq 0,\ \alpha+\beta = 1,\ \textit{if } \lambda \in \langle 2\pi/n, 2\pi/(n-1)\rangle,$$

or

$$\lambda^n = \alpha + \beta\lambda^q, \qquad \alpha \geq 0,\ \beta \geq 0,\ \alpha+\beta = 1,\ \textit{if } \lambda \in \langle 2\pi/q, 4\pi/n\rangle.\text{(9)}$$

λ cannot be in the angle $\langle 2\pi/(n-1), 2\pi/q\rangle$.(10)

PROOF. Let the stochastic matrix A admit λ as a characteristic root. By Theorem 5.3, in the corresponding Markov chain the system can go from $\mathscr{E}_i$ to either $\mathscr{E}_{i+1}$ or $\mathscr{E}_{i+2}$ in one unit of time. Therefore the probability σ_m of returning to the initial state after m units of time is zero if $2m < n$.

Thus

$$\left.\begin{aligned} n\sigma_m &= \sum_{i=1}^n \lambda_i^m = 0 \quad (m \leq q), \\ n\sigma_m &= \sum_{i=1}^n \lambda_i^m \geq 0 \quad (q+1 \leq m \leq n). \end{aligned}\right\} \tag{1}$$

We write the characteristic equation of the matrix as

$$\lambda^n = a_1\lambda^{n-1} + a_2\lambda^{n-2} + \cdots + a_n. \tag{2}$$

The coefficients $a_1, \dots, a_n$ are related to $s_m = n\sigma_m = \sum_{i=1}^n \lambda_i^m$ by the Newton recursive formula:

$$s_m = a_1 s_{m-1} + a_2 s_{m-2} + \cdots + a_{m-1}s_1 + a_m \cdot m \qquad (m = 1, 2, \dots, n).$$

From these formulas it is seen that

$$\left.\begin{aligned} a_m &= 0 \qquad (m \leq q), \\ a_m &= \tfrac{1}{m}s_m \quad (q+1 \leq m \leq n). \end{aligned}\right\} \tag{3}$$

(8) *Editor's note.* It is necessary that $n = 2q+1$ for the arguments used in the proof to hold. The incorrect definition of n has led to a few errors in the statement and proof of this theorem. Corrections are given.

(9) This should be: if $\lambda \in \langle 2\pi/(q+1), 4\pi/n\rangle$.

(10) This should be: λ cannot be in the angle $\langle 2\pi/(n-1), 2\pi/(q+1)\rangle$.

Equation (2) then becomes

$$\lambda^n = a_{q+1}\lambda^q + a_{q+2}\lambda^{q-1} + \cdots + a_n. \tag{4}$$

The elements a_{ij} of an arbitrary stochastic matrix satisfy the equation $\sum_{j=1}^n a_{ij} = 1$ $(i = 1, \dots, n)$. This implies that 1 is a characteristic root of any stochastic matrix, and in particular, a root of (4):

$$1 = a_{q+1} + a_{q+2} + \cdots + a_n. \tag{5}$$

We denote by Q the convex hull of the points $1, \lambda, \lambda^2, \dots, \lambda^{n-1}$. By (3), (4), and (5), $\lambda Q \subset Q$. Hence Q is an n-gon since $\lambda \notin M_{n-1}$. Clearly Q contains the points λ^n, $\lambda^{n+1}, \dots$, and it is the cyclic n-gon generated by λ.

We now suppose that $\lambda \in \langle 2\pi/p, 2\pi/(p-1)\rangle$, where $q+1 \le p \le n$.(11) Then the vertices of Q, in counterclockwise order, are

$$1, \lambda^p, \lambda, \lambda^{p+1}, \lambda^2, \dots, \lambda^{n-p-1}, \lambda^{n-1}, \lambda^{n-p}, \lambda^{n-p+1}, \dots, \lambda^{p-1}.$$

By Theorem 5.1,

$$\lambda^n = \lambda \cdot \lambda^{n-1} = \alpha\lambda^{n-p} + \beta\lambda^{n-p+1}, \qquad \alpha, \beta \ge 0, \quad \alpha + \beta = 1;$$

that is,

$$\lambda^p = \alpha + \beta\lambda, \qquad \alpha, \beta \ge 0, \quad \alpha + \beta = 1.$$

Consequently, the convex hull of the points $1, \lambda, \dots, \lambda^{p-1}$ is mapped into itself through multiplication by λ, which is only possible when $p \ge n$, and since $p \le n$, we must have $p = n$.

Thus we have proved that

1°. $\lambda \notin \langle 2\pi/(n-1), 2\pi/q\rangle$,(12) and

2°. if $\lambda \in \langle 2\pi/n, 2\pi/(n-1)\rangle$ then $\lambda^n = \alpha + \beta\lambda$, $\alpha, \beta \ge 0$, $\alpha + \beta = 1$.

It remains to consider the case where $\lambda \in \langle 2\pi/q, 4\pi/n\rangle$.(13) In this case, the vertices of Q, in counterclockwise order, are $1, \lambda^{q+1}, \lambda, \dots, \lambda^{n-1}, \lambda^q$. By Theorem 5.1,

$$\lambda^n = \lambda \cdot \lambda^{n-1} = \alpha + \beta\lambda^q, \qquad \alpha, \beta \ge 0, \quad \alpha + \beta = 1.$$

17. THEOREM VI. *If λ is an extremal number belonging to the intersection of $M_n - M_{n-1}$ with $\langle 2\pi/n, 2\pi/(n-1)\rangle$ and if $n = 2q$ is an even number, then λ generates a cyclic n-gon and satisfies the relation $\lambda^n = \alpha + \beta\lambda$, $\alpha, \beta \ge 0$, $\alpha + \beta = 1$.*

PROOF. Repeating the procedure used in §16, we find that the characteristic equation of A has the form

$$\left.\begin{aligned} &\lambda^n = a_q\lambda^q + a_{q+1}\lambda^{q-1} + \cdots + a_n, \\ &a_q \ge 0,\ a_{q+1} \ge 0, \dots, a_{n-1} \ge 0, \qquad a_q + a_{q+1} + \cdots + a_n = 1. \end{aligned}\right\} \tag{6}$$

(11)This should be $q+2 \le p \le n$.

(12)This should be $\lambda \notin \langle 2\pi/(n-1), 2\pi/(q+1)\rangle$.

(13)This should be $\lambda \in \langle 2\pi/(q+1), 4\pi/n\rangle$.

The difference from the case where n is odd lies in the fact that for $\lambda \in \langle 2\pi/n, 4\pi/n\rangle$ the coefficient a_n can be negative. We show, however, that in this case $\lambda \in \langle 2\pi/n, 2\pi/(n-1)\rangle$ and $a_n \geq 0$.

Let $a_n < 0$. Then (6) can be written in the form

$$b_1\lambda^n + b_2 = c_1\lambda + c_2\lambda^2 + \cdots + c_q\lambda^q,$$

$$b_1 > 0,\ b_2 \geq 0,\ b_1 + b_2 = 1,\ c_i \geq 0 \quad (i = 1, \ldots, q), \quad \sum_1^q c_i = 1.$$

This shows that the line segment $(1, \lambda^n]$ intersects the convex hull P of $\lambda, \lambda^2, \ldots, \lambda^{n-1}$, by the condition $\lambda \in \langle 2\pi/n, 2\pi/(n-1)\rangle$, $\lambda^n \in \langle 0, \arg\lambda\rangle$; but none of the vertices of P fall within this angle (see Figure 10). Therefore if the segment $(1, \lambda^n]$ meets P, then $\lambda^n \in P$. But if

$$\begin{aligned} &\lambda^n = d_1\lambda + d_2\lambda^2 + \cdots + d_{n-1}\lambda^{n-1}, \\ &d_1 \geq 0,\ d_2 \geq 0, \ldots, d_{n-1} \geq 0, \\ &d_1 + d_2 + \cdots + d_{n-1} = 1, \end{aligned}$$

or

$$\lambda^{n-1} = d_1 + d_2\lambda + \cdots + d_{n-1}\lambda^{n-2},$$

then the convex hull of the points $1, \lambda, \ldots, \lambda^{n-2}$ is mapped into itself through multiplication by λ, which is impossible since $\lambda \notin M_{n-1}$. Thus $a_n \geq 0$ and the proof can be completed as in §16.

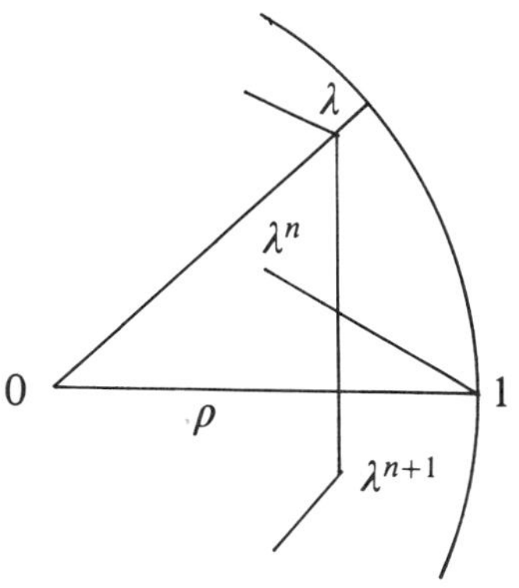

FIGURE 10

§4. Examples

18. In this section we construct M_n, $n \leq 5$. M_n is symmetric with respect to the real axis, since if λ is a characteristic root of the matrix A, then so is its complex conjugate $\overline{\lambda}$. Therefore it suffices to describe that part of M_n in the angle $\langle 0, \pi\rangle$.

As already noted, $M_2 = [-1, +1]$. From Theorems IV and V it follows that M_3 consists of the segment $[-1, +1]$ and the regular triangle with vertices $1, e^{2\pi i/3}$,

$e^{4\pi i/3}$. Furthermore, Theorems IV, V, and VI enable one to construct the intersection of M_4 with $\langle 0, 2\pi/3\rangle$ and the intersection of M_5 with $\langle 0, 4\pi/5\rangle$. We shall define the remaining part of M_4 lying in $\langle 2\pi/3, \pi\rangle$, and the remaining part of M_5 lying in $\langle 4\pi/5, \pi\rangle$.

19. Let the extremal number λ belong to the intersection of $M_4 - M_3$ with $\langle 2\pi/3, \pi\rangle$. Also let λ be a characteristic root of A. By Theorem 5.2 the matrix A is of the form

$$A = \begin{pmatrix} 0 & \alpha_1 & \beta_1 & 0 \\ 0 & 0 & \alpha_2 & \beta_2 \\ \beta_3 & 0 & 0 & \alpha_3 \\ \alpha_4 & \beta_4 & 0 & 0 \end{pmatrix}, \qquad \alpha_i \geq 0,\ \beta_i \geq 0,\ \alpha_i + \beta_i = 1.$$

The characteristic equation of A is

$$\lambda^4 = f_0 + f_1\lambda + f_2\lambda^2, \tag{7}$$

where $f_0 = \alpha_1\alpha_2\alpha_3\alpha_4 - \beta_1\beta_2\beta_3\beta_4$, $f_1 \geq 0$, and $f_2 = \beta_1\beta_3 + \beta_2\beta_4$. Note that

$$f_2^2 + 4f_0 = (\beta_1\beta_3 - \beta_2\beta_4)^2 + 4\alpha_1\alpha_2\alpha_3\alpha_4 \geq 0.$$

Put $f_2^2 + 4f_0 = 4c^2$. We express f_0 in terms of f_2 and c and substitute in (7), giving

$$(\lambda^2 - a)^2 = b^2\lambda + c^2 \tag{8}$$

where $a = \frac{1}{2}f_2$ and $b = \sqrt{f_1}$. We remark that

$$0 \leq a \leq 1, \qquad b > 0, \qquad (1-a)^2 = b^2 + c^2. \tag{9}$$

Now put $\lambda = \mu^2$, with $\frac{4}{3}\pi \leq \arg\mu \leq \frac{3}{2}\pi$; then([14]) $\arg\mu \leq \arg\mu^4 = 4\arg\mu - 4\pi$ and consequently

$$\mu^4 = \alpha + \beta\mu, \qquad \alpha \geq 0, \quad \beta \geq 0. \tag{10}$$

Substituting $\lambda = \mu^2$ and $\mu^4 = \alpha + \beta\mu$ in (8), we obtain a quadratic equation in μ:

$$(\alpha - a + \beta\mu)^2 = b^2\mu^2 + c^2.$$

Since μ is complex, the discriminant of this equation is negative (or zero if the equation reduces to an identity), which leads to

$$\left(\frac{\alpha - a}{c}\right)^2 + \left(\frac{\beta}{b}\right)^2 \leq 1.$$

Put $(\alpha - a)/c = R\cos\varphi$ and $\beta/b = R\sin\varphi$, $R \leq 1$. Using (9) and the Schwarz inequality, we obtain

$$\alpha + \beta = a + R(c\cdot\cos\varphi + b\cdot\sin\varphi) \leq a + R\sqrt{b^2+c^2} = a + R(1-a) \leq a + (1-a) = 1. \tag{11}$$

(10) and (11) imply that μ generates a cyclic quadrilateral Q. Clearly $\lambda Q = \mu^2 Q \subset Q$.

([14]) $\arg\mu$ denotes the value of the argument of μ lying between 0 and 2π.

If $\alpha+\beta < 1$, then the vertex μ^2, after multiplication by $\lambda = \mu^2$, will fall inside Q, contradicting Theorem 5.1. So $\alpha + \beta = 1$. Substituting $\mu = \sqrt{\lambda}$ in (10) and eliminating the square root, we obtain the following equation defining λ:

$$(\lambda^2 - \alpha)^2 = \beta^2\lambda, \qquad \alpha, \beta \geq 0, \quad \alpha + \beta = 1.$$

20. The case where λ is an extremal number belonging to the intersection of $M_5 - M_4$ with $\langle 4\pi/5, \pi\rangle$ carries over analogously. If λ is a characteristic root of A, the characteristic equation is of the form

$$\lambda^5 = f_0 + f_1\lambda + f_3\lambda^3, \qquad f_0 \geq 0, \qquad f_3^2 + 4f_1 \geq 0. \tag{12}$$

Equation (12) can be rewritten as

$$\lambda(\lambda^2 - a)^2 = b^2 + c^2\lambda, \tag{13}$$

where

$$a = \frac{1}{2}f_3, \qquad b = \sqrt{f_0}, \qquad c = \frac{1}{2}\sqrt{f_3^2 + 4f_1}.$$

Substituting $\lambda = \mu^2$ in (13), $\mu \in \langle 2\pi/5, \pi/2\rangle$, we obtain

$$(\mu^5 - a\mu)^2 = b^2 + c^2\mu^2. \tag{14}$$

Recalling that

$$\mu^5 = \alpha + \beta\mu, \qquad \alpha, \beta \geq 0, \tag{15}$$

we substitute the expression for μ in (15) into (14):

$$(\alpha + (\beta - a)\mu)^2 = b^2 + c^2\mu^2.$$

As above we observe that $\alpha + \beta = 1$. Thus μ generates a cyclic pentagon and λ is a vertex of this pentagon. From (15) we get the following equation defining λ:

$$\lambda(\lambda^2 - \beta)^2 = \alpha^2, \qquad \alpha, \beta \geq 0, \quad \alpha + \beta = 1.$$

21. The results of this section can be summarized in the following theorem.

THEOREM VII. *For $n \leq 5$, M_n is the union of all cyclic k-gons ($k \leq n$).*

Figure 11 represents M_2, M_3, M_4, M_5 and those parts of M_n, $n > 5$, which are known to us.

Appendix

All the previous results can easily be extended from stochastic matrices to the case of arbitrary matrices with nonnegative elements.

THEOREM VIII. (15) *An irreducible*(16) *matrix $A = (a_{ij})_1^n$ with nonnegative elements is similar*(17) *to the matrix ρB, where ρ is the spectral radius of A and B is a stochastic matrix of order n.*

PROOF. By Perron's theorem [3], ρ is a characteristic root of A and all the coordinates of the corresponding characteristic vector $(x_1, \ldots, x_n)$ are nonneg-

(15) Communicated by A. N. Kolmogorov.

(16) A matrix $A = (a_{ij})_1^n$ is called *reducible* if there exist subsets $p_1, \ldots, p_r$ and $q_1, \ldots, q_s$ of the numbers $1, \ldots, n$ such that $a_{p_\alpha q_\beta} = 0$ for $\alpha = 1, \ldots, r$ and $\beta = 1, \ldots, s$.

(17) I.e., represents the same linear mapping into another system of coordinates, consequently has the same characteristic roots.

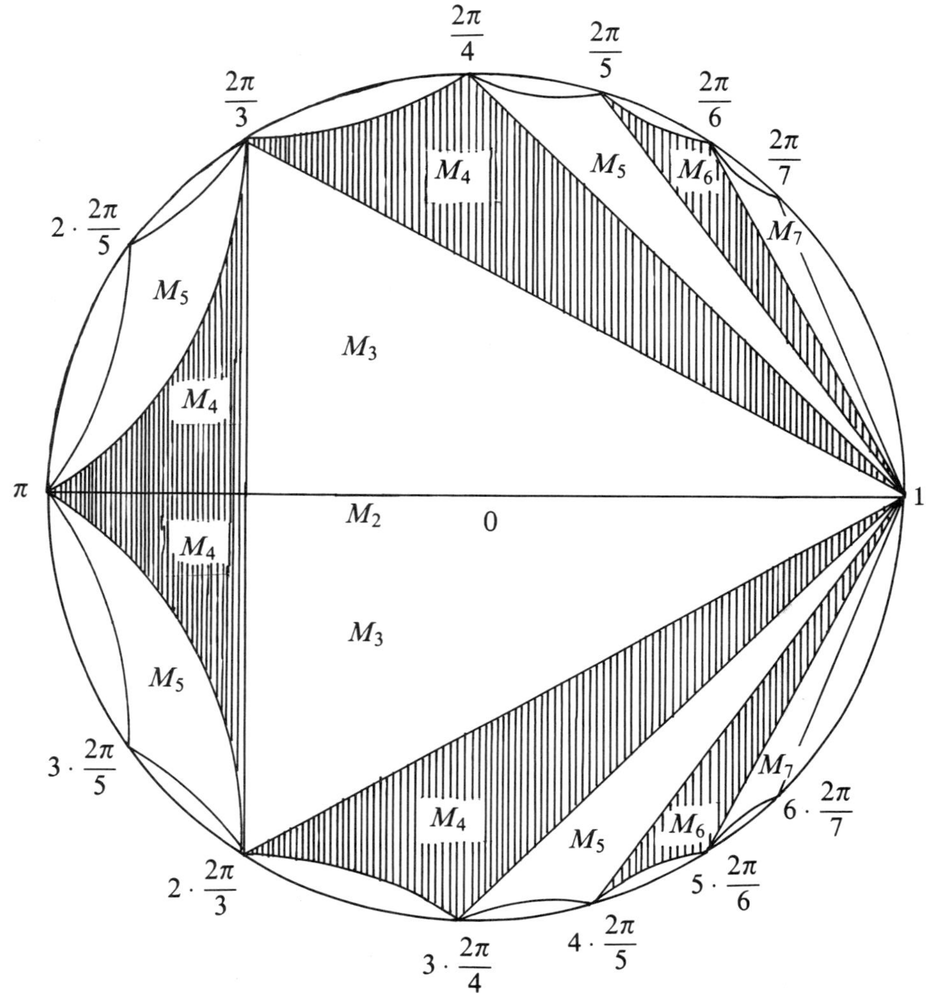

FIGURE 11

ative. They are strictly positive since the matrix A is irreducible. In fact, if $x_i = 0$ for $i = p_1, \ldots, p_r$ and $x_i > 0$ for $i = q_1, \ldots, q_s$, then by the identities $\sum_{j=1}^{n} a_{ij} x_j = \rho x_i$ we would have that $a_{p_\alpha q_\beta} = 0$ for $\alpha = 1, \ldots, r$, $\beta = 1, \ldots, s$.

The matrix C with elements $c_{ij} = a_{ij} \cdot x_j / x_i$ is similar to A. In fact $C = X^{-1}AX$, where

$$X = \begin{pmatrix} x_1 & 0 & \ldots & 0 \\ 0 & x_2 & \ldots & 0 \\ \ldots & \ldots & \ldots & \ldots \\ 0 & 0 & \ldots & x_n \end{pmatrix};$$

moreover, $c_{ij} \geq 0$ and

$$\sum_{j=1}^{n} c_{ij} = \sum_{j=1}^{n} a_{ij} x_j / x_i = \rho.$$

Hence C is equal to ρB, where B is a stochastic matrix.

THEOREM IX. *The characteristic roots of the matrix A of order n with non-negative elements belong to the region ρM_n where M_n is the region studied in sections 1 to 4, and ρ is the spectral radius of A.*

PROOF. For irreducible matrices, Theorem IX follows directly from Theorem VIII. If A is reducible, then by permutation of rows and columns A can be put in the form

$$\begin{pmatrix} A_{11} & 0 & 0 & \dots & 0 \\ A_{21} & A_{22} & 0 & \dots & 0 \\ \dots & \dots & \dots & \dots & \dots \\ A_{k1} & A_{k2} & A_{k3} & \dots & A_{kk} \end{pmatrix}$$

where $A_{11}, A_{22}, \dots, A_{kk}$ are irreducible (or zero). Thus each characteristic root of A is a characteristic root of one of the diagonal matrices $A_{11}, \dots, A_{kk}$, which implies that it belongs to one of the regions $\rho_\nu M_{e_\nu}$ where e_ν is the order of $A_{\nu\nu}$ and ρ_ν the spectral radius of $A_{\nu\nu}$. It follows then that all the characteristic roots of A must lie in the region ρM_n, where ρ is the largest of the ρ_ν's.

1. V. Romanovsky [V. I. Romanovskiĭ], *Recherches sur les chaînes de Markoff. Premier mémoire*, Acta Math. **66** (1936), 147–251.

2. N. Dmitriev and E. Dynkin, *On the characteristic numbers of a stochastic matrix*, C. R. (Dokl.) Acad. Sci. URSS **49** (1945), 159–163.

3. Oskar Perron, *Grundlagen für eine Theorie des Jacobischen Kettenbruchalgorithmus*, Math. Ann. **64** (1907), 1–76.

Addendum
Root Location for Stochastic Matrices

George Phillip Barker

Let M_n be the set of all complex numbers z such that z is an eigenvalue of some $n \times n$ stochastic matrix. The problem of characterizing M_n was posed by A. N. Kolmogorov (cf. [2]) and partially solved by Dmitriev and Dynkin [3]. The set M_n is a subset of the closed unit disc and is mapped into itself upon multiplying by certain complex numbers λ. They are thus led to introduce a class of convex polygons which they call cyclic. They were able to determine M_n completely for $n \leq 5$ and partially for $n > 5$.

In 1951 Karpelevich [7] was able to completely describe the sets M_n. He also defined a cyclic convex polygon, but his definition is different from that of Dmitriev and Dynkin. A discussion of these differences together with nice illustrations of both types of cyclic polygons and of some of the sets M_n can be found in the McGill University M. Sc. thesis of Joanne Swift [15].

The sets M_n of course settle the inverse problem for a single complex number, and it is natural to consider the full inverse problem: what are necessary and sufficient conditions on a set of n complex numbers that they should be the spectrum of some stochastic, or more generally nonnegative, matrix? It is in the latter form that the bulk of research has focused.

However, Suleĭmanova [13] began with the inverse problem for stochastic matrices and considered sets which contain a single real number of modulus 1. Other sufficient conditions are given. A translation of this paper appears as Appendix B of Swift's thesis [15], where this paper and the subsequent work of Perfect [9]–[11] are discussed. Another discussion of these ideas can be found in Suleĭmanova's later paper [14].

A discussion of Suleĭmanova's condition is given by Friedland in [5], where he notes that if $\lambda_1, \ldots, \lambda_n$ is the spectrum of a nonnegative matrix, then necessarily for $k = 1, 2, \ldots$ we have

$$\sum_{j=1}^{n} \lambda_j^k \geq 0. \tag{$*$}$$

He poses the following conjecture.

Let $\alpha = \{\lambda_1, \ldots, \lambda_n\}$ be a set of complex numbers. Assume that α satisfies $(*)$ for $k = 1, 2, \ldots$. If α contains exactly one positive number then α is the spectrum of some nonnegative matrix. He then gives various results which are supporting evidence for this conjecture.

Another natural restriction on sets of possible eigenvalues is that they all be real. This is discussed by Salzmann [12] who, like Suleĭmanova, gives conditions on a set of real numbers that they be the spectrum of a nonnegative diagonalizable matrix. If the nonnegative matrix is to have real spectrum and be diagonalizable, then when may it be chosen to be symmetric? This form of the inverse problem was posed and solved by Fiedler [4]. A discussion of Fiedler's solution and other aspects of symmetric nonnegative matrices is given in Chapter 8 of Berman and Plemmon's book [1]. See also the notes to that chapter for further references.

Finally note that a different sort of restriction can be imposed. If A is nonnegative, then it is well known that we may associate a directed graph with A (cf. Varga [16] or Berman and Plemmons [1]). What is the set of all possible eigenvalues of a nonnegative matrix with special graph? Some results on this problem can be found in [6] and [8].

References

1. Abraham Berman and Robert J. Plemmons, *Nonnegative matrices in the mathematical sciences*, Academic Press, 1979.

2. N. Dmitriev and E. Dynkin, *On the characteristic numbers of a stochastic matrix*, C. R. (Dokl.) Acad. Sci. URSS **49** (1945), 159–163.

3. ——, *Characteristic roots of stochastic matrices*, Izv. Akad. Nauk SSSR Ser. Mat. **10** (1946), 167–184; English transl. in this volume.

4. M. Fiedler, *Eigenvalues of nonnegative symmetric matrices*, Linear Algebra and Appl. **9** (1974), 119–142.

5. Shmuel Friedland, *On an inverse problem for nonnegative and eventually nonnegative matrices*, Israel J. Math. **29** (1978), 43–60.

6. Charles R. Johnson, R. B. Kellogg, and A. B. Stephens, *Complex eigenvalues of a nonnegative matrix with a specified graph*. II, Linear and Multilinear Algebra **7** (1979), 129–143; **8** (1979/80), 171.

7. F. I. Karpelevich, *On the characteristic roots of matrices with nonnegative elements*, Izv. Akad. Nauk SSSR Ser. Mat. **15** (1951), 361–383; English transl. in this volume.

8. R. B. Kellogg and A. B. Stephens, *Complex eigenvalues of a nonnegative matrix with a specified graph*, Linear Algebra and Appl. **20** (1978), 179–187.

9. Hazel Perfect, *On positive stochastic matrices with real characteristic roots*, Proc. Cambridge Philos. Soc. **48** (1952), 271–276.

10. ——, *Methods of constructing certain stochastic matrices*, Duke Math. J. **20** (1953), 395–404.

11. ——, *Methods of constructing certain stochastic matrices*. II, Duke Math. J. **22** (1955), 305–311.

12. Frank L. Salzmann, *A note on eigenvalues of nonnegative matrices*, Linear Algebra and Appl. **5** (1972), 329–338.

13. Kh. R. Suleĭmanova, *Stochastic matrices with real characteristic numbers*, Dokl. Akad. Nauk SSSR **66** (1949), 343–345. (Russian)

14. ——, *On the question of a necessary and sufficient condition for the existence of a stochastic matrix with prescribed characteristic numbers*, Trudy Vsesoyuz. Zaochn. Ènerget. Inst. Vyp. **28** (1965), 33–49. (Russian)

15. J. Swift, *The location of characteristic roots of stochastic matrices*, M. Sc. thesis, McGill Univ., Montréal, 1972.

16. Richard S. Varga, *Matrix iterative analysis*, Prentice-Hall, Englewood Cliffs, N.J., 1962.

Translated by GEORGE FLEISCHER

Amer. Math. Soc. Transl.
(2) Vol. **140**, 1988

On the Characteristic Roots of Matrices with Nonnegative Elements

F. I. KARPELEVICH

Abstract. This paper considers the distribution of the characteristic roots of the nth order matrices with nonnegative elements and a fixed maximum modulus of the characteristic roots.

§1.

Consider the set $\mathfrak{M}_n$ of all nth order matrices with nonnegative elements such that the maximum modulus of their characteristic roots is equal to ρ, and let M_n^ρ denote the set of characteristic roots of all such matrices. The set M_n^ρ, with the usual geometric interpretation of complex numbers, can be represented by some plane figure. Our problem is to determine the figure M_n^ρ. This problem was posed by A. N. Kolmogorov in 1938 in connection with the theory of Markov chains and was partially solved in 1946 by Dmitriev and Dynkin [1], who calculated the part of the figure M_n^ρ included within a certain angle and, in addition, completely determined M_n^ρ for $n \leq 5$.

In the present paper Kolmogorov's problem will be completely solved.

It is obvious that M_n^ρ can be obtained from M_n^1 by a homothety from the point O with coefficient ρ; hence it suffices to find M_n^1 (which we will henceforth denote by M_n, omitting the superscript).

A convex n-gon Q will be called *cyclic* if there exist a complex number λ and an integer $p \leq n$ such that Q is the convex hull of the system of points

$$\lambda^m e^{2\pi i q/p} \quad (m = 0, 1, \ldots; q = 0, 1, \ldots, p-1).$$

Our main results are the following two theorems.

THEOREM A. *The figure M_n is the set-theoretic union of all cyclic k-gons for $k \leq n$.*

1980 *Mathematics Subject Classification* (1985 *Revision*). Primary 15A18, 15A48.
Translation of Izv. Akad. Nauk SSSR Ser. Mat. **15** (1951), 361–383; MR **13**, 201.

Theorem A was stated as a conjecture in [1]. However, by a cyclic polygon was meant only a polygon that is the convex hull of the points $1, \lambda, \lambda^2, \dots$.

Despite the fact that we will find explicit equations of the curves forming the boundary of M_n (see Theorem B), the question of whether Theorem A is true under this narrower meaning of the term "cyclic polygon" remains open.

THEOREM B. *The figure M_n is symmetric with respect to the real axis, is contained in the disc $|z| \le 1$, and has the points $e^{2\pi i a/b}$ $(0 \le a < b \le n)$ in common with the circle $|z| = 1$. The boundary of M_n consists of these points and curvilinear arcs joining them in circular order. For $n > 3$, each of these arcs is defined by one of the following parametric equations:*

$$\text{I.}\quad \lambda^q(\lambda^p - t)^r = (1-t)^r,$$
$$\text{II.}\quad (\lambda^b - t)^d = (1-t)^d\lambda^q,$$

where the parameter t varies over the interval $0 \le t \le 1$, and $b, d, p, q,$ and r are natural numbers defined as follows.

Suppose the endpoints of some arc taken counterclockwise are $e^{2\pi i a'/b'}$ and $e^{2\pi i a''/b''}$. There are two cases:

$$\text{a)}\quad b''[n/b''] \ge b'[n/b'],$$
$$\text{b)}\quad b''[n/b''] \le b'[n/b'].$$

If a) *holds for some arc, then* b) *holds for the complex conjugate arc, and conversely. Therefore, in view of the symmetry of M_n, it suffices to determine the arcs satisfying condition* a).

Suppose $r_1 = b''$, $r_2 = a''$, $r_3, \dots, r_m$ is the sequence of remainders obtained in the process of finding the greatest common divisor of b'' and a'' by means of the Euclidean algorithm. If $[n/b''] = 1$ and $r_{2s} = 1$ for some integer s, then the arc joining the points $e^{2\pi i a'/b'}$ and $e^{2\pi i a''/b''}$ is given by equation I, *where $r = r_{2s-1}$ and the numbers p and q are determined from the relations*

$$a''p \equiv 1 \bmod b'' \quad (0 < p < b''),$$
$$a''q \equiv -r \bmod b'' \quad (0 \le q < b'').$$

Otherwise, the arc joining the points $e^{2\pi i a'/b'}$ and $e^{2\pi i a''/b''}$ is given by equation II, *where $d = [n/b'']$, $b = b''$, and q is determined from the relation*

$$a''q \equiv -1 \bmod b'' \quad (0 < q < b'').$$

As an example, in the table we give the equations of the boundary of M_7. In view of the symmetry of M_n, it suffices to give the equations for the part of the boundary lying in the half-plane $\operatorname{Im} z \geq 0$.

a'/b'	a''/b''	Equation of the part of the boundary joining $e^{2\pi i a'/b'}$ and $e^{2\pi i a''/b''}$	Type of the equation
0	1/7	$(\lambda - t)^7 = (1-t)^7$	I
1/7	1/6	$\lambda^7 - t = (1-t)\lambda$	II
1/6	1/5	$\lambda^6 - t = (1-t)\lambda$	II
1/5	1/4	$\lambda^5 - t = (1-t)\lambda$	II
1/4	2/7	$\lambda^7 - t = (1-t)\lambda^3$	II
2/7	1/3	$\lambda(\lambda^3 - t)^3 = (1-t)^3$	I
1/3	2/5	$(\lambda^3 - t)^2 = (1-t)^2\lambda$	II
2/5	3/7	$\lambda^7 - t = (1-t)\lambda^2$	II
3/7	1/2	$\lambda(\lambda^2 - t)^3 = (1-t)^3$	I

§2.

According to a theorem of A. N. Kolmogorov (see the Appendix to [1]), the figure M_n coincides with the set of characteristic roots of all nth order stochastic matrices, i.e., those matrices (a_{ij}) such that $a_{ij} \geq 0$ $(i,j = 1,\ldots,n)$ and $\sum_{j=1}^n a_{ij} = 1$ $(i = 1,\ldots,n)$.

Using this theorem, Dmitriev and Dynkin showed that $\lambda \in M_n$ if and only if there exists in the complex plane a convex k-gon Q $(k \leq n)$, not consisting of zero alone, such that $\lambda Q \subset Q$. It follows that M_n is star-shaped, i.e., if λ belongs to M_n, then so does the whole radius joining λ to the point O. Moreover, it is easy to see that M_n is closed; hence the boundary of M_n comprises those numbers λ such that $\lambda \in M_n$, but $\rho\lambda \notin M_n$ for any $\rho > 1$. Following [1], we will call such numbers λ *extremal.* We will require the main result of [1]:

DMITRIEV–DYNKIN THEOREM. *Suppose $\lambda \in M_n - M_{n-1}$(1) is an extremal number and $2\pi(k-1)/n \leq \arg\lambda \leq 2\pi k/n$ $(0 < k < n)$. If a convex polygon Q contains λQ and if $x_0, x_1, \ldots, x_{n-1}$ are the vertices of Q, enumerated in order when proceeding counterclockwise around the contour of Q, then*

$$\lambda x_s = \alpha_s x_{s+k-1} + \beta_s x_{s+k}, \qquad \alpha_s, \beta_s \geq 0, \quad \alpha_s + \beta_s = 1$$

(*the indices are regarded modulo n*).

Moreover, it is not possible to have some $\alpha_i = 0$ and some $\beta_j = 0$ simultaneously. We will always assume that all $\beta_j > 0$. If, for some extremal number λ and some polygon Q, all $\beta_j > 0$ and some $\alpha_i = 0$, then, for the extremal number $\overline{\lambda}$ and the polygon $\overline{Q}$, some $\beta_j = 0$ and all $\alpha_i > 0$. Therefore the condition $\beta_j > 0$ $(j = 0, 1, \ldots, n-1)$ distinguishes a part T_n of the boundary of the figure $M_n - M_{n-1}$ such that the entire boundary of $M_n - M_{n-1}$ is the set-theoretic sum of T_n and $\overline{T}_n$.

(1)The inclusion $M_{n-1} \subset M_n$ is obvious.

§3.

Let us fix an extremal number $\lambda \in T_n$ such that $2\pi(k-1)/n \le \arg\lambda \le 2\pi k/n$ $(0 < k < n)$ and consider, without mentioning it in each particular instance, only convex n-gons Q such that $\lambda Q \subset Q$.

LEMMA 1. *Suppose $x_0, x_1, \dots, x_{n-1}$ are the vertices of a polygon Q, taken in cyclic order proceeding counterclockwise around its contour, and suppose $\lambda x_s = x_{s+k}$. Then the polygon Q' with vertices $x_0, x_1, \dots, x_{s+k-2}, \lambda x_{s-1}, x_{s+k}, x_{s+k+1}, \dots, x_{n-1}$ is sent into part of itself under multiplication by λ.*

Indeed, it is easy to see (Figure 1) that $\lambda Q \subset Q' \subset Q$. Therefore $\lambda Q' \subset \lambda Q \subset Q'$. According to the Dmitriev–Dynkin theorem, the vertices of Q are sent onto the contour of Q under multiplication by λ. Some of these vertices will hit vertices of Q, while others will hit interior points of sides. Our first problem will be to study the relative arrangement of these two types of vertices.

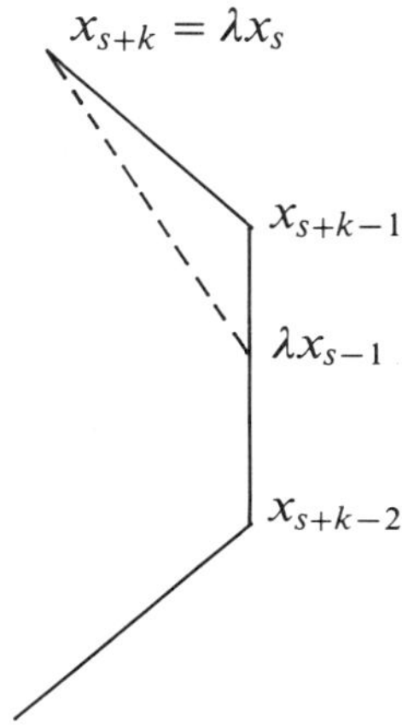

FIGURE 1

We will schematize and simplify the problem before us by introducing the following game. The board consists of n playing spaces arranged in a circle. On some of these spaces stand chips. We are allowed to move any chip across $k-1$ spaces to the kth counterclockwise, provided that the space immediately ahead of the chip is unoccupied. If as a result of some move a chip falls on a space occupied by another, the chip occupying that space is removed from the board. Consider an arbitrary polygon Q such that $\lambda Q \subset Q$, and associate in cyclic order the vertices of Q with the spaces of our board. Place a chip on each space corresponding to a vertex of Q that is not sent into a vertex of Q (i.e., that is sent into an interior point of a side) under multiplication by λ. The configurations so constructed for all possible polygons Q will be called *admissible*. It follows from Lemma 1 that configurations obtained from an admissible configuration in accordance with the rules of our game are again admissible. Our immediate goal is to select from all admissible configurations the simplest one possible.

To solve this problem we need a lemma from number theory. Let us denote by $R_n(m)$ the remainder obtained when m is divided by n. Fix an integer k such that $0 < k \leq n$ and consider the sequence

$$a_0 = 0,\ a_1, \ldots, a_m, \ldots, \tag{A}$$

where $a_m = R_n(mk)$, and also the sequence

$$r_1 = n,\ r_2 = k,\ r_3, \ldots, r_s \geq 0 \tag{R}$$

formed by the remainders obtained in the course of determining the greatest common divisor $d = D(n,k)$ of n and k by means of the Euclidean algorithm. Let us also put $r_0 = n+k$ and $r_{-1} = 2n+k$. We will frequently use sequences (A) and (R), and we will adhere to the same notation for their elements throughout this paper.

We call a_m a *minimal term* of sequence (A) if $0 < a_m < a_t$ for $t = 1, \ldots, m-1$; we call a_m a *maximal term* of sequence (A) if $a_m > a_t$ for $t = 0, 1, \ldots, m-1$.

There are a finite number of minimal and maximal terms. We denote them by $\mu_0, \mu_1, \ldots, \mu_m$ in the order in which they appear in the sequence (A).

LEMMA 2.

1. *If μ_i is a minimal term of sequence* (A) *and $r_{2p} \leq \mu_i < r_{2p-2}$ for some p, then $\mu_i = vr_{2p-1} + r_{2p}$, where v is a nonnegative integer.*

1*. *If μ_i is a maximal term of sequence* (A) *and $r_{2p+1} \leq n - \mu_i < r_{2p-1}$ for some p, then $n - \mu_i = vr_{2p} + r_{2p+1}$, where v is a nonnegative integer.*

2. *If μ_i is a minimal term of sequence* (A) *and $r_{2p} \leq \mu_i < r_{2p-2}$ for some p, then*

a) $\max_{0 \leq j < i} \mu_j = n - r_{2p-1}$,

b) $\min_{0 < j < i} \mu_j = \mu_i + r_{2p-1}$.

2*. *If μ_i is a maximal term of sequence* (A) *and $r_{2p+1} \leq n - \mu_i < r_{2p-1}$ for some p, then*

a) $\min_{0 < j \leq i} \mu_j = r_{2p}$,

b) $\max_{0 \leq j < i} \mu_j = \mu_i - r_{2p}$.

We will prove Lemma 2 by induction. For $i \leq 2$ it can easily be verified directly. Now suppose $i \geq 2$ and Lemma 2 is true for all μ_j with $0 \leq j \leq i$; we will prove it for μ_{i+1}. We consider two cases:

(A) μ_i is a minimal term;

(B) μ_i is a maximal term.

Case (A). Obviously, $r_{2p} \leq \mu_i < r_{2p-2}$ for some p. We divide case (A) into two subcases:

(A_1) $r_{2p} < \mu_i < r_{2p-2}$,

(A_2) $r_{2p} = \mu_i$.

(A_1). Suppose $\mu_i = a_q$. We choose from the sequence $a_0, a_1, \ldots, a_{q-1}$ the largest term $\nu_1 = a_m$. It follows from induction assumption 2a) that $a_m = \nu_1 = n - r_{2p-1}$. From the sequence $a_0, a_1, \ldots, a_{m-1}$ we choose the largest term ν_2.

It follows from induction assumption 2*b) that $\nu_2 = n - r_{2p-1} - r_{2p-2}$. By induction assumption 1, $\mu_i = vr_{2p-1} + r_{2p}$ and, since $r_{2p} < \mu_i$, we have $v \geq 1$.

Consider the sequence $\mu_i = a_q, a_{q+1}, a_{q+2}, \dots, a_{q+m}$. It is obvious that

$$a_q = R_n(a_q + a_0),\ a_{q+1} = R_n(a_q + a_1), \dots, a_{m+q} = R_n(a_q + a_m).$$

Since $0 < a_t \leq \nu_2$ for $0 < t < m$, we have

$$\mu_i = a_q < a_q + a_t \leq \mu_i + n - r_{2p-1} - r_{2p-2} < n - r_{2p-1} = \nu_1 < n \quad (t = 1, \dots, m-1).$$

Therefore

$$\mu_i < R_n(a_q + a_t) = a_q + a_t < \nu_1.$$

This shows that among the terms $a_{q+1}, a_{q+2}, \dots, a_{q+m-1}$ there are neither minimal nor maximal terms of sequence (A).

We have

$$a_q + a_m = \mu_i + \nu_1 = vr_{2p-1} + r_{2p} + n - r_{2p-1} = (v-1)r_{2p-1} + r_{2p} + n;$$

since $v \geq 1$, it follows that $(v-1)r_{2p-1} + r_{2p} \geq r_{2p} \geq 0$, and therefore

$$a_{q+m} = R_n(a_q + a_m) = (v-1)r_{2p-1} + r_{2p} < vr_{2p-1} + r_{2p} = \mu_i.$$

This proves that

$$a_{q+m} = (v-1)r_{2p-1} + r_{2p} = \mu_i - r_{2p-1} = \mu_{i+1}$$

is a minimal term of sequence (A), and now it is easy to verify that all parts of Lemma 2 are satisfied for μ_{i+1}.

(A_2). Suppose $\mu_i = a_q$. Choose $\nu_1 = a_m$ and ν_2 as in Case (A_1). As before, among the terms $a_{q+1}, \dots, a_{q+m-1}$ there are neither minimal nor maximal terms of sequence (A).

We have

$$a_{q+m} = R_n(a_q + a_m) = R_n(\mu_i + \nu_1) = R_n(r_{2p} + n - r_{2p-1}) = n - r_{2p-1} + r_{2p};$$

suppose $r_{2p-1} = ur_{2p} + r_{2p+1}$ (by definition of a minimal term, $0 < a_q = \mu_i = r_{2p}$, and so division is possible). Then

$$\nu_1 = n - r_{2p-1} < a_{q+m} = n - r_{2p-1} + r_{2p} = n - (u-1)r_{2p} - r_{2p+1}.$$

Since $i \geq 2$, it follows that $q \geq 2$ and therefore $\nu_1 = a_m \geq a_1$; hence $m \geq 1$ and $a_{q+m} \neq a_q$. Thus we have proved that

$$a_{q+m} = n - (u-1)r_{2p} - r_{2p+1} = \mu_{i+1}$$

is a maximal term of sequence (A), and now the proof of Lemma 2 presents no difficulty.

Case (B) is handled analogously.

COROLLARY 1. a) *For any* $m \geq 1$, r_{2m} *is a minimal term of sequence* (A) *if* $r_{2m} \neq 0$.

b) *For any* $m \geq 1$, $n - r_{2m-1}$ *is a maximal term of sequence* (A).

For the proof, note that each a_i is divisible by d; hence $d \leq a_i \leq n - d$ if $a_i \neq 0$. Therefore d is a minimal term and $n - d$ a maximal term of sequence (A).

It is now easy to prove Corollary 1 by applying parts 2a) and 2*a) of Lemma 2 in succession.

Let us turn to the study of our configurations. We divide the chips of an arbitrary configuration into groups, putting two chips A and B into the same group if and only if there exists a sequence of chips $A_1, \ldots, A_m$ such that the pairs of chips (A, A_1), (A_i, A_{i+1}), $i = 1, \ldots, m-1$, and (A_m, B) are situated on adjacent spaces.

We denote by $\varphi(\Gamma)$ the number of chips in the configuration Γ, and by $\psi(\Gamma)$ the number of groups in Γ. If a group does not occupy the whole board, then the space directly ahead of the group is free; hence the leading chip in this group can in one move be shifted k spaces ahead, after which the second chip in the group can be shifted, and so on; as a result of a series of such moves, the whole group is shifted k spaces ahead. Thus we can state

LEMMA 3. *If there are free spaces on the board, then by a series of moves it is possible to shift any group of chips as a whole* k *spaces ahead, without moving the chips of the other groups.*

THEOREM I. *Any configuration* Γ *can be transformed in a series of steps into a configuration* Γ_0 *such that either* $\psi(\Gamma_0) = 1$ *and* $n - \varphi(\Gamma_0)$ *is a maximal term of sequence* (A), *or* $\varphi(\Gamma_0) < d$.

PROOF. We consider the set of configurations obtained from Γ in a finite number of steps, and distinguish those with the smallest number of groups; from among those configurations we choose a configuration Γ_0 having the smallest number of chips.

a) We will prove that either $\psi(\Gamma_0) = 1$ or $\varphi(\Gamma_0) < d$.

We consider some group of chips G and distinguish d consecutive spaces on the board such that the first chip of G stands on the first of the distinguished spaces. Suppose $\psi(\Gamma_0) > 1$. Then there exists a group G_1 different from G. By Lemma 3, it can be shifted k spaces ahead; in view of the choice of Γ_0, the two groups cannot merge, and therefore the shifted group G_1 can be shifted again k spaces ahead, and so on. We shift the group G_1 until the last of the chips of G_1 falls on one of the distinguished spaces. All other chips of G_1 will precede this last chip and, at the same time, will follow the group G; hence they will also fall on distinguished spaces. In the same way, we can situate the other groups on these spaces; hence $\varphi(\Gamma_0) < d$.

b) We will now prove that if $\varphi(\Gamma_0) \geq d$, then $n - \varphi(\Gamma_0)$ is a maximal term of sequence (A).

Suppose $\varphi(\Gamma_0) \geq d$; then, in view of a), $\psi(\Gamma_0) = 1$. Our assertion is satisfied if $\varphi(\Gamma_0) = n$. Consider the case where $\varphi(\Gamma_0) < n$. Number the spaces of the board in order counterclockwise so that those occupied by chips receive the numbers $n-\varphi+1, n-\varphi+2, \ldots, n-1, 0$, where $\varphi = \varphi(\Gamma_0)$. Under this numbering space 1 is free; hence the chip standing on space 0 can be moved to space a_1, then to space a_2, and so on. We move this chip as long as possible. We will show that under this series of moves the chip visits one of the spaces $n-\varphi, \ldots, n-1$. Indeed, the chip can stop only on one of these spaces. On the other hand, if it passed infinitely many times, it would visit all spaces with numbers divisible by d, and since there are at least d of the spaces $n-\varphi, \ldots, n-1$, at least one of these numbers must be divisible by d. Note, however, that in its circuit the chip cannot hit any of the spaces $n-\varphi+1, \ldots, n-1$, since in this case the number of chips would decrease, while the number of groups would remain equal to 1. Therefore the chip stops on space $n-\varphi$, without visiting any of the spaces $n-\varphi+1, \ldots, n-1$. But this means that $n-\varphi$ is a maximal term of sequence (A).

THEOREM II. *If $\lambda^n \neq 1$, then there exists a polygon Q_0 such that $\lambda Q_0 \subset Q_0$ and*

a) *all vertices of Q_0 that do not pass into vertices of Q_0 under multiplication by λ are situated consecutively;*

b) *if φ is the number of such vertices, then $\varphi \geq d$ and $n-\varphi$ is a maximal term of sequence* (A).

PROOF. By Lemma 1, a configuration obtained from an admissible one by means of a single move is again admissible; hence, by Theorem I, there exists an admissible configuration Γ_0 such that either

a) $\psi(\Gamma_0) = 1$, or

b) $n-\varphi(\Gamma_0)$ is a minimal term of sequence (A) or $\varphi(\Gamma_0) < d$.

Suppose the configuration Γ_0 corresponds to a polygon Q_0. We will show that if $\varphi(\Gamma_0) < d$, then $\lambda^n = 1$. Suppose $x_0, x_1, \ldots, x_{n-1}$ are the vertices of Q_0, numbered in order proceeding counterclockwise around the contour. We assign to the vertex x_i an index j, where $i \equiv j \bmod d$ $(0 \leq j < d)$. Since there are fewer than d vertices not sent into vertices under multiplication by λ, there must be an index j such that all vertices with this index are sent into vertices under multiplication by λ. Consider any vertex x corresponding to the index j. Since multiplication by λ corresponds to increasing the number of the vertex by k, multiplication by λ does not alter the index of a vertex. Therefore all of the points x, $\lambda x, \ldots, \lambda^n x$ are vertices with index j; obviously $\lambda^n x = x$, and hence $\lambda^n = 1$. We will henceforth assume $\lambda^n \neq 1$.

Let $x_0, x_1, \ldots, x_{n-1}$ be the vertices of Q_0, numbered in order counterclockwise. Suppose the vertices $x_{n-k+1}, \ldots, x_{n-k+\varphi}$ are not sent into vertices of Q_0 under multiplication by λ. Put

$$\xi_i = \lambda x_{n-k+i}, \quad i = 0, 1, \ldots, n-1$$

(the indices are regarded modulo n). By the Dmitriev–Dynkin theorem,

$$\xi_i = \alpha_i x_{i-1} + \beta_i x_i, \qquad \alpha_i + \beta_i = 1, \quad \alpha_i \geq 0,\ \beta_i \geq 0,\ i = 0, 1, \ldots, n-1. \quad (1)$$

Obviously for any vertex x of Q_0 there exists a unique natural number c such that $\lambda^c x = \xi_i$, $0 < i \leq \varphi$. In particular, this applies to the vertices $x_0, \ldots, x_\varphi$. Our immediate problem is to determine how the vertices $x_0, \ldots, x_\varphi$ pass into the points $\xi_1, \ldots, \xi_\varphi$ under multiplication by a power of λ. We will consider separately two cases: $\varphi > d$ and $\varphi = d$.

THEOREM III. *Suppose $n - \nu$ is a maximal term of sequence* (A) *and $\nu \geq \varphi > d$. Then*

$$\begin{aligned} \lambda^p x_i &= \xi_{i+r_{2m}}, & i &= 0, 1, \ldots, \nu - r_{2m}, \\ \lambda^{p+q} x_i &= \xi_{i+r_{2m}} - \nu, & i &= \nu - r_{2m} + 1, \ldots, \nu, \end{aligned} \qquad (2)$$

where the numbers m, p, and q are uniquely determined from the conditions

$$0 < r_{2m} < \nu \leq r_{2m-1}, \qquad kp \equiv r_{2m} \bmod n, \qquad 0 \leq p < n/d,$$
$$kq \equiv -\nu \bmod n, \qquad 0 \leq q < n/d.$$

We will first show that for some m

$$\text{(a)} \quad r_{2m} < \nu \leq r_{2m-1}.$$

By Lemma 2, for some m we have

$$r_{2m+1} \leq \nu = v r_{2m} + r_{2m+1} < r_{2m-1}.$$

If $v \neq 0$ and $r_{2m+1} > 0$, then $r_{2m} < \nu$ and (a) is proved; if $v \neq 0$ but $r_{2m+1} = 0$, then $r_{2m} = d < \varphi \leq \nu$, which proves (a). Finally, if $v = 0$, then $\nu = r_{2m+1} \geq \varphi > d$; hence $r_{2m+2} > 0$ and therefore

$$\nu = r_{2m+1} = u r_{2m+2} + r_{2m+3};$$

it follows that $r_{2(m+1)} < \nu = r_{2(m+1)-1}$ and (a) is satisfied if we replace m by $m+1$. It follows from (a) that $r_{2m-1} \geq \nu > d$, and hence $r_{2m} > 0$.

Multiplication by λ corresponds to increasing the index of a vertex by k. Since the vertices of Q_0 that are not sent into vertices under multiplication by λ can all be found among $x_{n-k+1}, \ldots, x_{n-k+\nu}$, and since a vertex that is sent into the point $\xi_{i+r_{2m}}$ under multiplication by λ is $x_{n-k+i+r_{2m}}$, it follows that for $\lambda^p x_i = \xi_{i+r_{2m}}$ it is sufficient that in the sequence

$$R_n(i + a_0),\ R_n(i + a_1), \ldots, R_n(i + a_t), \ldots$$

there appears the term $n - k + i + r_{2m}$, and not later than any of the numbers $n-k+1, \ldots, n-k+\nu$. Obviously this will happen if and only if in the sequence

$$R_n(a_1 + i),\ R_n(a_2 + i), \ldots, R_n(a_t + i), \ldots$$

there appears the term $R_n(r_{2m} + i)$, and not later than any of the numbers $1, \ldots, \nu$.

By Corollary 1 of Lemma 2, r_{2m} is a minimal term of sequence (A); hence if $a_p = r_{2m}$, then by Lemma 2, part 2a),

$$a_j \leq n - r_{2m-1} \quad (j = 0, \ldots, p-1).$$

Consequently,

$$a_j + i \leq n - r_{2m-1} + \nu - r_{2m} \leq n - r_{2m} < n \quad \text{for} \quad 0 \leq j < p,\ i \leq \nu - r_{2m}. \tag{3}$$

It follows from Lemma 2, part 2b), that

$$a_j \geq r_{2m} + r_{2m-1} \quad (j = 1, \ldots, p-1),$$

and hence

$$a_j + i \geq r_{2m} + r_{2m-1} + i > \nu \quad \text{for} \quad 0 < j < p,\ i > -r_{2m}. \tag{4}$$

It follows from (3) and (4) that

$$\nu < a_j + i < n,$$
$$i = -r_{2m} + 1, \ldots, 0, \ldots, \nu - r_{2m}, \quad j = 1, \ldots, p-1.$$

Thus

$$\nu < R_n(a_j + i),$$
$$i = n - r_{2m} + 1, \ldots, 0, \ldots, \nu - r_{2m}, \quad j = 1, \ldots, p-1.$$

This shows that in each sequence

$$R_n(a_1 + i),\ R_n(a_2 + i), \ldots, R_n(a_p + i)$$
$$(i = n - r_{2m} + 1, \ldots, 0, \ldots, \nu - r_{2m})$$

the term $R_n(r_{2m} + i) = R_n(a_p + i)$ occurs, and no later than any of the numbers $1, \ldots, \nu$. Therefore

$$\lambda^p x_i = \xi_{i + r_{2m}} \quad (i = n - r_{2m} + 1, \ldots, 0, \ldots, \nu - r_{2m}). \tag{2a}$$

We will now show that for some $q \geq 0$

$$\lambda^q x_i = x_{n - \nu + i} \quad (i = \nu - r_{2m} + 1, \nu - r_{2m} + 2, \ldots, \nu). \tag{2b}$$

Theorem III will be completely proved by the equalities (2a) and (2b).

As before, in order that the equalities (2b) be satisfied it is sufficient that in the sequence $R_n(a_1 + i), \ldots, R_n(a_t + i), \ldots$ there occurs the term $R_n(n + i - \nu)$, and not later than any of the numbers $1, \ldots, \nu$. By hypothesis, $n - \nu$ is a maximal term of sequence (A); hence for some $q \geq 0$ we have $a_q = n - \nu$ and therefore $R_n(a_q + i) = R_n(n - \nu + i)$. We will show that none of $1, 2, \ldots, \nu$ occurs among the numbers

$$R_n(a_1 + i),\ R_n(a_2 + i), \ldots, R_n(a_{q-1} + i) \quad (i = \nu - r_{2m} + 1, \ldots, \nu).$$

It follows from Lemma 2, parts 2*a) and 2*b), that

$$r_{2m} \leq a_j \leq n - (r_{2m} + \nu), \quad j = 1, \ldots, q-1.$$

Therefore

$$a_j + i \leq n - r_{2m} < n \quad \text{and} \quad a_j + i \geq r_{2m} + \nu - r_{2m} + 1 > \nu$$
$$(i = \nu - r_{2m} + 1, \ldots, \nu; \quad j = 1, \ldots, q-1),$$

which proves everything.

Since a_p is a minimal term of sequence (A), p is the smallest positive integer such that $kp \equiv r_{2m} \bmod n$. Note, however, that sequence (A) is periodic with period n/d. Therefore, the smallest positive solution p of the congruence $kp \equiv r_{2m} \bmod n$ satisfies $0 \leq p < n/d$ and is uniquely determined by this inequality. Analogously, $kq \equiv -\nu \bmod n$, $0 \leq q < n/d$.

COROLLARY I. *According to Theorem* II, *if* $\varphi > d$, *then* $n - \varphi$ *is a maximal term of sequence* (A); *hence we can put* $\nu = \varphi$ *in the equalities* (2), *thereby obtaining*

$$\begin{aligned} \lambda^p x_i &= \xi_{i+r_{2m}}, & i &= 0, 1, \ldots, \varphi - r_{2m}, \\ \lambda^{p+q} x_i &= \xi_{i+r_{2m}-\varphi} & i &= \varphi - r_{2m} + 1, \ldots, \varphi, \end{aligned} \tag{5}$$

where $0 < r_{2m} < \varphi \leq r_{2m-1}$, $kp \equiv r_{2m} \bmod n$ $(0 \leq p < n/d)$, *and* $kq \equiv -\varphi \bmod n$ $(0 \leq q < n/d)$.

THEOREM IV. *Suppose* $\varphi = d$. *Then*

$$\lambda^l x_i = \xi_i, \quad i = 1, \ldots, d, \tag{6}$$

where $l = n/d$.

PROOF. As in Theorem III, the equalities (6) will be satisfied if in the sequence $R_n(a_1 + i), \ldots, R_n(a_t + i), \ldots$ there occurs the number i, and no later than any of the numbers $1, \ldots, d$.

Let $n/d = l$. Then $a_l = 0$ and $a_j > 0$ for $0 < j < l$. Since any a_j must be divisible by d, it follows that $d \leq a_j \leq n - d$ if $a_j \neq 0$; therefore $d \leq a_j \leq n - d$ $(0 < j < l)$. Thus

$$a_j + i \geq d + 1 \quad \text{and} \quad a_j + 1 \leq n \quad (i = 1, \ldots, d; j = 1, \ldots, l-1).$$

It follows that none of the numbers $1, \ldots, d$ occurs among $R_n(a_1 + i)$, $R_n(a_2 + i), \ldots, R_n(a_{l-1} + i)$ $(i = 1, \ldots, d)$. But $R_n(a_l + i) = R_n(i) = i$, which proves the theorem.

Since the vertices of Q_0 that do not pass into vertices under multiplication by λ are all accounted for by $x_{n-k+1}, \ldots, x_{n-k+\varphi}$, the only vertices of Q_0 that do not pass into vertices under division by λ are $x_1, \ldots, x_\varphi$. Therefore all of the vertices of Q_0 have the form $\lambda^t x_i$ $(i = 1, \ldots, \varphi)$ and, for any t, $\lambda^t x_i \neq x_j$ if $i \neq j$ and $1 \leq i, j \leq \varphi$. It is clear from (5) that when $\varphi > d$ it suffices to give t the values $(0, \ldots, p-1)$ if $1 \leq i \leq \varphi - r_{2m}$ and the values $(0, 1, \ldots, p+q-1)$ if $\varphi - r_{2m} < i \leq \varphi$. Thus if $\varphi > d$, then all vertices of Q_0 are accounted for by the vertices

$$\lambda^t x_i \quad \begin{matrix} t = 0, \ldots, p-1, \\ i = 1, \ldots, \varphi - r_{2m}, \end{matrix} \qquad \lambda^t x_i \quad \begin{matrix} t = 0, 1, \ldots, p+q-1, \\ i = \varphi - r_{2m} + 1, \ldots, \varphi. \end{matrix} \tag{7}$$

Analogously, it follows from (6) that if $\varphi = d$, all vertices of Q_0 are accounted for by the vertices

$$\lambda^t x_i \quad \begin{array}{l} t = 0, 1, \dots, l-1, \\ i = 1, \dots, d. \end{array} \tag{8}$$

The vertices of the form $\lambda^t x_i$ will be called the vertices generated by the vertex x_i. It follows from (7) and (8) that any vertex of Q_0 is generated by exactly one of the vertices $x_1, \dots, x_\varphi$.

§4.

THEOREM V. *If for the polygon Q_0 we have $\varphi > d$ and $0 < r_{2m} < \varphi \leq r_{2m-1}$, then $r_{2m} = 1$ if $n > 3$.*

Assume $r_{2m} > 1$. Our problem is to modify Q_0 so that the modified polygon is sent into part of itself under multiplication by λ and at least one of its vertices is sent inside Q_0. We will thus obtain a contradiction to the Dmitriev–Dynkin theorem.

Consider the polynomial path Y with vertices

$$y_i = \frac{x_i}{\lambda^{p+q}} \quad (i = 0, \dots, r_{2m}) \quad \text{and} \quad y_i = \frac{x_i}{\lambda^p} \quad (i = r_{2m}+1, \dots, \varphi). \tag{9}$$

We denote by δ_i the interval (x_{i-1}, x_i) and by Δ_i the interval (y_{i-1}, y_i). The equalities (9) yield

$$\begin{aligned} \lambda^{p+q}\Delta_i &= \delta_i \quad (i = 1, \dots, r_{2m}), \\ \lambda^p \Delta_i &= \delta_i \quad (i = r_{2m}+2, \dots, \varphi). \end{aligned} \tag{10}$$

It follows from (1) and (5) that

$$\begin{aligned} \lambda^p x_i &= \xi_{i+r_{2m}} \in \delta_{i+r_{2m}} && (i = 0, 1, \dots, \varphi - r_{2m}), \\ \lambda^{p+q} x_i &= \xi_{i+r_{2m}-\varphi} \in \delta_{i+r_{2m}-\varphi} && (i = \varphi - r_{2m}+1, \dots, \varphi), \end{aligned}$$

and hence, in view of (10),

$$\begin{aligned} x_i &\in \frac{\delta_{i+r_{2m}}}{\lambda^p} = \Delta_{i+r_{2m}} && (i = 2, \dots, \varphi - r_{2m}), \\ x_i &\in \frac{\delta_{i+r_{2m}-\varphi}}{\lambda^{p+q}} = \Delta_{i+r_{2m}-\varphi} && (i = \varphi - r_{2m}+1, \dots, \varphi). \end{aligned} \tag{11}$$

The equalities (11) show that the vertices $x_2, \dots, x_\varphi$ lie on the sides of Y. The rest of the proof is divided into two cases:

1) $2r_{2m} \leq \varphi + 1$;

2) $2r_{2m} > \varphi + 1$.

1) In this case we will alter the vertices $x_1, \dots, x_{r_{2m}}$. On the line x_0x_1 we take a point x_1' and project it through the point ξ_2 on the side of Y passing through x_2 (see Figure 2); we project the resulting point x_2' through the point ξ_3 on the side of Y passing through x_3, and so on, until we obtain the point $x_{r_{2m}}'$ (see Figure 3). We then join the points $x_{r_{2m}}'$ and $x_{r_{2m}+1}$ by a straight line.

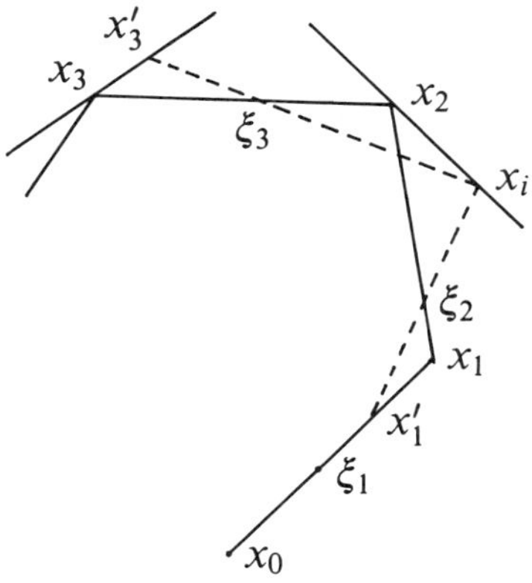

FIGURE 2

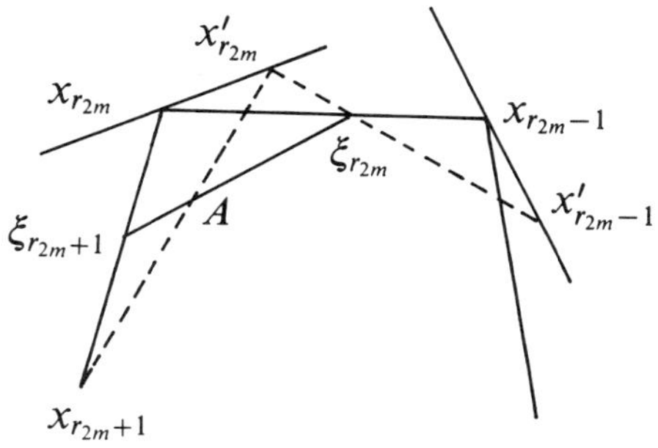

FIGURE 3

Since, by assumption, $r_{2m} > 1$, it follows that $x'_{r_{2m}}$ does not coincide with x'_1.

Consider the polygon Q obtained from Q_0 as follows: the vertices of the form $\lambda^t x_i$ $(i = 1, \dots, r_{2m})$ are replaced by the vertices $\lambda^t x'_i$; the remaining vertices are unchanged. It is obvious that the polygon Q is completely determined as soon as the point x'_1 is chosen. We will show that x'_1 can be chosen so that $\lambda Q \subset Q$ and one of the vertices of Q leaves the contour of Q under multiplication by λ.

Indeed, we will explain how the polygon λQ is situated. The vertices of Q are divided into two groups:

1) the vertices that are also vertices of Q_0, and

2) the vertices generated by $x'_1, \dots, x'_{r_{2m}}$.

Therefore if we replace $x_1, \dots, x_{r_{2m}}$ in (7) by $x'_1, \dots, x'_{r_{2m}}$, the resulting vertices will account for all vertices of Q. It follows that the vertices of Q are sent into vertices of Q under multiplication by λ, except for the vertices

$$\lambda^{p-1} x'_i \quad (i = 1, \dots, r_{2m}), \qquad \lambda^{p-1} x_i \quad (i = r_{2m} + 1, \dots, \varphi - r_{2m}),$$
$$\lambda^{p+q-1} x_i \quad (i = \varphi - r_{2m} + 1, \dots, \varphi)$$

if $2r_{2m} < \varphi + 1$, and the vertices

$$\lambda^{p-1} x'_i \quad (i = 1, \dots, r_{2m} - 1), \quad \lambda^{p+q-1} x'_{r_{2m}}, \quad \lambda^{p+q-1} x_i \quad (i = r_{2m} + 1, \dots, \varphi)$$

if $2r_{2m} = \varphi + 1$.

Thus in order that $\lambda Q \subset Q$ it is necessary and sufficient that Q contain the points

$$\lambda^p x'_i \quad (i = 1, \dots, r_{2m}), \qquad \lambda^p x_i \quad (i = r_{2m} + 1, \dots, \varphi - r_{2m}),$$
$$\lambda^{p+q} x_i \quad (i = \varphi - r_{2m} + 1, \dots, \varphi)$$

if $2r_{2m} < \varphi + 1$, and the points

$$\lambda^p x'_i \quad (i = 1, \dots, r_{2m} - 1), \qquad \lambda^{p+q} x'_{r_{2m}}, \qquad \lambda^{p+q} x_i \quad (i = r_{2m} + 1, \dots, \varphi)$$

if $2r_{2m} = \varphi + 1$.

If the point x_1' is taken sufficiently close to x_1 so that the point ξ_1 turns out to be inside Q, then Q will obviously contain all the ξ_i $(i = 1, \dots, \varphi)$, and, in view of (5), all vertices in group 1) will be sent into the contour of Q under multiplication by λ.

Suppose $2r_{2m} < \varphi + 1$. In this case we see from (11) that

$$x_i \in \Delta_{i+r_{2m}} \quad (i = 2, 3, \dots, r_{2m}).$$

If x_1' is taken close to x_1, then obviously

$$x_i' \in \Delta_{i+r_{2m}} \quad (i = 2, \dots, r_{2m}).$$

In view of (10), we obtain

$$\lambda^p x_i' \in \delta_{i+r_{2m}}.$$

It is easy to see, however, that the vertices x_i $(i = r_{2m} + 1, \dots, 2r_{2m})$, which are endpoints of $\delta_{i+r_{2m}}$ $(i = 2, \dots, r_{2m})$, are vertices of Q belonging to the first group. Therefore if $2r_{2m} < \varphi + 1$, the points $\lambda^p x_1'$ $(i = 2, \dots, r_{2m})$ lie on the contour of Q.

Suppose $2r_{2m} = \varphi + 1$. As in the previous case, the points $\lambda^p x_i'$ $(i = 2, \dots, r_{2m} - 1)$ lie within the intervals $\delta_{i+r_{2m}}$ and therefore on the contour of Q.

It follows from (11) that

$$x_{r_{2m}} = x_{\varphi+1-r_{2m}} \in \Delta_1.$$

If the point x_1' is close to x_1, then $x'_{r_{2m}} \in \Delta_1$ and, in view of (10),

$$\lambda^{p+q} x'_{r_{2m}} \in \delta_1,$$

i.e., the point $\lambda^{p+q} x'_{r_{2m}}$ lies within the side $[x_0, x_1]$. If x_1' is close to x_1, then the point $\lambda^{p+q} x'_{r_{2m}}$ is close to the point $\lambda^{p+q} x_{r_{2m}} = \xi_1$ and therefore lies on the contour of Q.

Thus, both in the case $2r_{2m} < \varphi + 1$ and in the case $2r_{2m} = \varphi + 1$, in order that $\lambda Q \subset Q$ it is necessary and sufficient that x_1' be sufficiently close to x_1 and that Q contain the point $\lambda^p x_1'$.

We introduce coordinates on the lines $x_0 x_1$ and $\xi_{r_{2m}} \xi_{r_{2m}+1}$ by assigning to the points x_1 and $\xi_{r_{2m}+1}$ the coordinate 0 and to the points x_0 and $\xi_{r_{2m}}$ the coordinate 1.

Let A be the point of intersection of the line $x'_{r_{2m}} x_{r_{2m}+1}$ and the line $\xi_{r_{2m}} \xi_{r_{2m}+1}$ (see Figure 3); let x be the coordinate of the point x_1' on the line $x_0 x_1$, and $u(x)$ the coordinate of the point A on the line $\xi_{r_{2m}} \xi_{r_{2m}+1}$. Since $\lambda^p x_0 = \xi_{r_{2m}}$ and $\lambda^p x_1 = \xi_{r_{2m}+1}$ (Theorem III), any point on $x_0 x_1$ is sent into the point on $\xi_{r_{2m}} \xi_{r_{2m}+1}$ with the same coordinate under multiplication by λ^p. Therefore, in order that the point $\lambda^p x_1'$ be inside Q it is necessary and sufficient that $x > u(x)$. If, in addition, x is sufficiently small, i.e., x_1' is close to x_1, then $\lambda Q \subset Q$.

Since the point A was obtained from x_1' by a series of projections, we obviously have

$$u(x) = (\tilde{a} x + \tilde{b}) / (\tilde{c} x + \tilde{d});$$

if x_1' coincides with x_1, then the polygon Q coincides with Q_0 and the points A with $\xi_{r_{2m}+1}$, i.e., $u(0) = 0$, and therefore

$$u(x) = \tilde{a}x/(\tilde{c}x + \tilde{d}).$$

Obviously $\tilde{d} \neq 0$; dividing the numerator and denominator by $\tilde{d}$, we obtain

$$u(x) = ax/(cx + 1) = ax(1 - cx + \cdots).$$

Therefore

$$x - u(x) = x(1 - a) + cx^2 + \cdots .$$

If $a \neq 1$, there exist numbers x arbitrarily close to 0 such that $x - u > 0$. Thus it remains to consider the case where $a = 1$ and $x - u = cx^2 + \cdots$.

Note that $0 < u(1) < 1$, as is evident from the drawing (see Figure 4). An exception is possible only if $n = 3$ (see Figure 5).(2) It follows that $c > 0$ and thus, for x sufficiently small, $x - u = cx^2 + \cdots > 0$.

2) $2r_{2m} > \varphi + 1$. In this case we will alter the vertices $x_{\varphi-1}, \ldots, x_{r_{2m}-1}$. The alteration is analogous to that of Case 1; it is illustrated in Figure 6. The analysis of this case is analogous to that of Case 1.

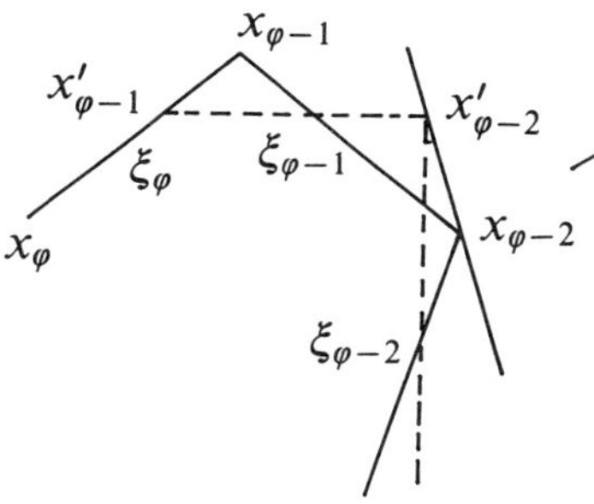

FIGURE 4

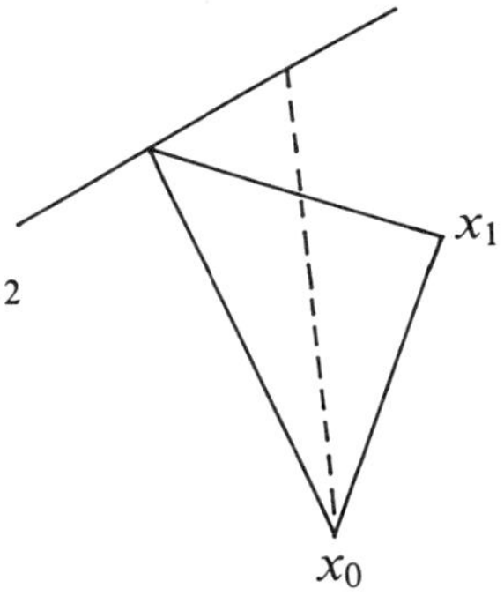

FIGURE 5

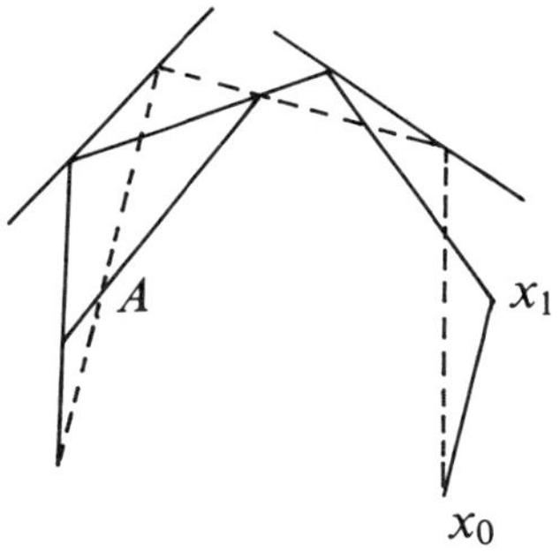

FIGURE 6

THEOREM VI. *If for the polygon Q_0 we have $\varphi > d$, then $d = 1$, and λ satisfies a relation*

$$\lambda^q(\lambda^p - \alpha)^r = \beta^r, \qquad \alpha, \beta \geq 0, \quad \alpha + \beta = 1, \tag{12}$$

where $r = r_{2m-1}$, $kp \equiv 1 \bmod n$ $(0 \leq p < n)$, $kq \equiv -r \bmod n$ $(0 \leq q < n)$, and m is defined by the condition $r_{2m} = d = 1$.

PROOF. For some m (see Corollary 1 of Theorem III) we have $0 < r_{2m} < \varphi \leq r_{2m-1}$. By Theorem V, $r_{2m} = 1$; hence $d = r_{2m} = 1$. By Corollary 1 of Lemma 2, $n - r_{2m-1}$ is a maximal term of sequence (A); hence we can put $\nu = r_{2m-1}$ in (2), thereby obtaining

$$\lambda^p x_i = \xi_{i+1} = \alpha_{i+1}x_i + \beta_{i+1}x_{i+1}, \qquad \lambda^{p+q}x_r = \xi_1 = \alpha_1 x_0 + \beta_1 x_1$$
$$(\alpha_i \geq 0,\ \beta_i \geq 0,\ \alpha_i + \beta_i = 1,\ i = 0, 1, \ldots, r-1), \tag{13}$$

(2)We will henceforth consider the case where $n > 3$. The figures M_n with $n \leq 3$ can be found in [1].

where $kp \equiv 1 \bmod n$ $(0 \le p < n)$ and $kq \equiv -r \bmod n$ $(0 \le q < n)$. We will show that $\alpha_1 = \alpha_2 = \cdots = \alpha_r = \alpha$ in (13). Consider the polygon Q obtained from Q_0 as follows: the vertices of the form $\lambda^t x_i$ are replaced by the vertices $\lambda^t[x_i + \varepsilon(x_i - x_{i-1})]$ $(1 \le i \le r-1)$; the other vertices are unchanged. If $\alpha_{i+1} \ne \alpha_i$, we can choose ε so that $\lambda Q \subset Q$ and one of the vertices of λQ falls inside Q. The proof of this assertion is analogous to that of Theorem IV in [1].

Thus we have $\alpha_1 = \cdots = \alpha_r = \alpha$ and, therefore, $\beta_1 = \cdots = \beta_r = \beta$. Consequently, (13) can be rewritten in the form

$$\lambda^p x_i = \alpha x_i + \beta x_{i+1}, \qquad \lambda^{p+q} x_r = \alpha x_0 + \beta x_1$$
$$(i = 0, 1, \ldots, r-1;\ \alpha \ge 0,\ \beta \ge 0,\ \alpha + \beta = 1).$$

Eliminating $x_0, x_1, \ldots, x_r$ from these equations, we obtain (12). To Q_0 there corresponds some nth order stochastic matrix (see [1]). Equation (12) is the characteristic equation of this matrix. Therefore equation (12) must be of degree n. Thus we have

$$q + pr = n. \tag{14}$$

THEOREM VII. *If for the polygon Q_0 we have $\varphi = d$, then λ satisfies a relation*

$$(\lambda^l - \beta)^d = \lambda^q \alpha^d, \quad \alpha \ge 0,\ \beta \ge 0,\ \alpha + \beta = 1, \tag{15}$$

where $l = n/d$, $kq \equiv -d \bmod n$, and $0 \le q < n/d$.

PROOF. Each vertex of Q_0 is generated by one of the vertices $x_1, \ldots, x_d$. Since multiplication by λ corresponds to increasing the index of a vertex by k, it is easy to see that the vertex generating x_0 is x_d. Consequently, there exists q such that $\lambda^q x_d = x_0$; hence $kq + d \equiv 0 \bmod n$. Furthermore, it is obvious that q is the smallest positive integer such that $kq \equiv -d \bmod n$. Therefore $0 \le q < n/d$.

We can show that $\alpha_1 = \cdots = \alpha_d = \alpha$ in the equalities (6) of §3. The proof is analogous to that of Theorem VI. Thus we have

$$\lambda^l x_i = \alpha x_{i-1} + \beta x_i \quad (i = 1, \ldots, d), \qquad \alpha, \beta \ge 0, \quad \alpha + \beta = 1,$$
$$\lambda^q x_d = x_0;$$

Eliminating $x_0, x_1, \ldots, x_d$, we obtain (15).

§5.

THEOREM VIII. *If $r_{2m} \ne 1$ for each m, then*

$$2\pi \frac{k-s}{n-q} < \arg \lambda < 2\pi \frac{k}{n}, \tag{16}$$

where s and q are uniquely determined from the relations

$$sn - qk = d, \qquad 0 \le q < l = n/d, \qquad 0 \le s < k/d.$$

PROOF. It follows from Theorems V and VII that there exists a polygon Q_0 with vertices $x_0, x_1, \ldots, x_{n-1}$, numbered in order counterclockwise, such that

$$x_{i-1} = \frac{\lambda^l - \beta}{\alpha} x_i \quad (i = 1, \ldots, d), \quad \alpha > 0, \quad \beta > 0, \quad \alpha + \beta = 1, \quad \lambda^q x_d = x_0. \tag{17}$$

Let $\delta_i = (x_i, x_{i-1})$, $i = 1, \ldots, d$. For any t we obviously have $\lambda^t x_i \neq x_j$ if $i \neq j$, $0 < i, j \leq d$; hence for any t

$$\lambda^l \delta_i \neq \delta_j \quad (i \neq j).$$

It follows from (17) that

$$\delta_i = \delta_d \left(\frac{\lambda^l - \beta}{\alpha}\right)^{d-i}. \tag{18}$$

Equality (18) shows that the angles at which the intervals δ_i are seen from the origin are equal. It is clear from (8) that the points

$$\lambda^t x_i \quad \begin{matrix} t = 0, 1, \ldots, l-1 \\ i = 1, \ldots, d \end{matrix}$$

are vertices of Q_0. Thus the intervals

$$\lambda^t \delta_1 \quad (t = 0, 1, \ldots, l-q-1),$$

$$\lambda^t \delta_i \quad \begin{matrix} t = 0, 1, \ldots, l-1 \\ i = 2, \ldots, d \end{matrix}$$

are sides of Q_0. Therefore, if α is the angle at which one of the intervals δ_i is seen from the origin, then

$$(l-q)\alpha + l(d-1)\alpha < 2\pi,$$

whence

$$\alpha < \frac{2\pi}{n-q}. \tag{19}$$

It follows from (17) that $\lambda^t x_1 \in \delta_1$; hence, as can be seen from Figure 7,

$$2\pi\left(1 - \frac{1}{n-q}\right) < 2\pi - \alpha < \arg \lambda^l < 2\pi. \tag{20}$$

On the other hand,

$$\frac{k}{n} - \frac{k-s}{n-q} = \frac{1}{l(n-q)} \quad \text{and} \quad (e^{2\pi i k/n})^l = 1,$$

whence

$$(e^{2\pi i(k-s)/(n-q)})^l = e^{2\pi i(n-q-1)/(n-q)}. \tag{21}$$

From (20) and (21) we easily obtain (16).

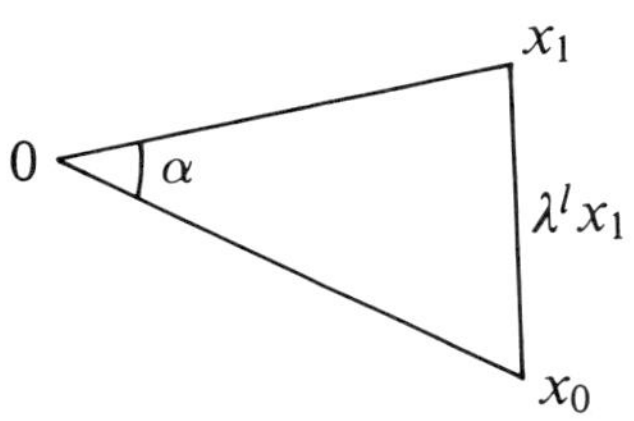

FIGURE 7

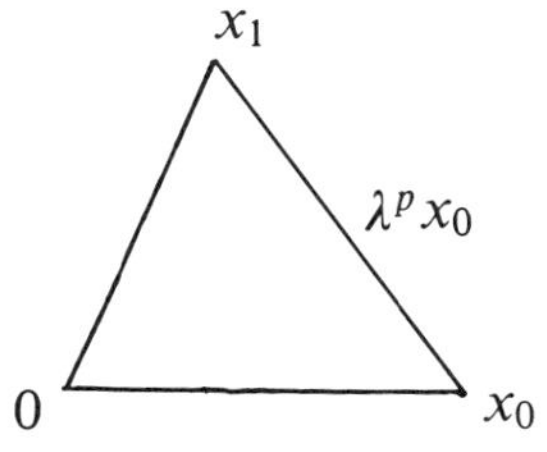

FIGURE 8

THEOREM IX. *If $r_{2m} = 1$ for some m, then*

$$2\pi \frac{s}{p} < \arg \lambda < 2\pi \frac{k}{n}, \tag{22}$$

where s and p are uniquely determined from the relation

$$kp - sn = 1, \quad 0 \leq p < n, \quad 0 \leq s < k.$$

PROOF. It follows from Theorem V that there exists a polygon Q_0 with vertices $x_0, x_1, \ldots, x_{n-1}$, numbered in order counterclockwise, such that

$$\lambda^p x_0 = \alpha x_0 + \beta x_1, \qquad \alpha, \beta > 0, \quad \alpha + \beta = 1.$$

This means that the point $\lambda^p x_0$ is situated within the side $[x_0, x_1]$. Therefore (see Figure 8)

$$0 < \arg \lambda^p < \pi. \tag{23}$$

But $\frac{k}{n} - \frac{s}{p} = \frac{1}{np}$ and $(e^{2\pi i s/p})^p = 1$; hence

$$(e^{2\pi i (k-1)/n})^p = e^{2\pi i (n-p+1)/n}.$$

In view of (14), we have $(n - p + 1)/n > 1/2$, which, together with (23), proves (22).

THEOREM X. *Each $\lambda \in T_n$ belongs to a cyclic n-gon.*

PROOF. It was shown in §4 that λ satisfies one of the relations (12) or (15). Suppose, for example, λ satisfies (12). Let δ be the greatest common divisor of q and r, and suppose $q = q'\delta$ and $r = r'\delta$. Putting $\mu^{r'} = \lambda$ in (12) and taking the rth root of both sides, we obtain

$$\mu^{q'}(\mu^{pr'} - \alpha) = \beta\gamma, \qquad \alpha \geq 0, \ \beta \geq 0, \quad \alpha + \beta = 1, \tag{24}$$

where $\gamma^r = 1$.

We will show that μ can be chosen so that $\gamma^\delta = 1$.

Indeed, if we put $\alpha = 0$ and $\beta = 1$ in (24), we obtain $\gamma = \mu^{q'+pr'}$. It follows from the definition of r' and q' and from (14) that $q' + pr' = n/\delta$; hence $\gamma = \mu^{n/\delta}$.

On the other hand, if we put $\alpha = 0$ and $\beta = 1$ in (12), we obtain $\lambda^n = 1$ and, in view of (22), $\lambda = e^{2\pi i k/n}$. Therefore,

$$\mu = \lambda^{1/r'} = e^{2\pi i (k+un)/(r'n)},$$

and hence

$$\gamma = e^{2\pi i (k+un)/(r'\delta)}.$$

It is easy to see that r' and n are relatively prime; hence there exists u such that $k + un \equiv 0 \bmod r'$, and so $\gamma^\delta = 1$. Choosing μ in (24) so that $\gamma^\delta = 1$, we obtain

$$\mu^{n/\delta} = \beta\gamma + \alpha u^{q'}, \qquad \alpha, \beta \geq 0, \quad \alpha + \beta = 1, \quad \gamma^\delta = 1. \tag{25}$$

It follows from (25) that the point $\mu^{n/\delta} e^{2\pi i \nu/\delta}$ is contained in the convex hull of the points

$$\mu^m e^{2\pi i \xi/\delta} \quad (m = 0, 1, \ldots, \tfrac{n}{\delta} - 1; \xi = 0, 1, \ldots, \delta - 1);$$

hence μ, and therefore also $\lambda = \mu^{r'}$, belongs to a cyclic n-gon.

We are now in a position to prove Theorem A. It follows from Theorem X that each $\mu \in M_n$ belongs to some cyclic m-gon, where $m \le n$. On the other hand, it is obvious that the product of two points of a cyclic polygon is again a point of this polygon, which proves Theorem A.

Let us turn to a proof of Theorem B. The statement of Theorem B was given in §1. We will prove it by induction. Theorem B is true for $n = 4$ (see [1]). Suppose Theorem B is true for every figure M_m, $4 \le m < n$.

a) We introduce some definitions and notation.

By the angle $\langle \varphi_1, \varphi_2 \rangle$ we mean the set of complex numbers λ such that $2\pi\varphi_1 < \arg \lambda < 2\pi\varphi_2$. By $\langle \overline{\varphi_1, \varphi_2} \rangle$ we mean the angle that is the complex conjugate of $\langle \varphi_1, \varphi_2 \rangle$.

It follows from Theorems VIII and IX that

$$\left\langle \frac{k-1}{n}, \frac{k}{n} \right\rangle \cap T_n \subset \left\langle \frac{f(k,n)}{\varphi(k,n)}, \frac{k}{n} \right\rangle, \tag{26}$$

where $\varphi(k,n)$ and $f(k,n)$ are determined by the conditions

$$k\varphi(k,n) - nf(k,n) = d, \quad 0 < k - f(k,n) < \frac{k}{d}, \quad 0 < n - \varphi(k,n) < \frac{n}{d}, \; 0 < k < n.$$

Fractions a_1/b_1 and a_2/b_2 ($a_1/b_1 < a_2/b_2$, $0 \le a_i < b_i \le n$, $i = 1,2$) will be called *adjacent* if there is no fraction a/b ($0 \le a < b \le n$) such that $a_1/b_1 < a/b < a_2/b_2$. If a_1/b_1 and a_2/b_2 are two fractions adjacent to k/n and if $a_1/b_1 < k/n < a_2/b_2$, then it follows from the theory of Farey series that

$$\left.\begin{array}{c} b_i k - a_i n = (-1)^{i+1} d, \\ 2k - a_1 - a_2 = \frac{k}{d}, \qquad 2n - b_1 - b_2 = \frac{n}{d}, \\ 0 < k - a_i < \frac{k}{d}, \qquad 0 < n - b_i < \frac{n}{d}, \quad i = 1,2. \end{array}\right\} \tag{27}$$

We conclude from (27) that $a_1 = f(k,n)$ and $b_1 = \varphi(k,n)$; hence the fractions $f(k,n)/\varphi(k,n)$ and k/n are adjacent.

b) We have

$$\left\langle \frac{k-1}{n}, \frac{k}{n} \right\rangle \cap \overline{T}_n \subset \left\langle \frac{k-1}{n}, \frac{a}{b} \right\rangle,$$

where the fractions $(k-1)/n$ and a/b are adjacent, $0 < k < n$.

Indeed, it follows from a) that

$$\left\langle \overline{\frac{k-1}{n}, \frac{k}{n}} \right\rangle \cap T_n = \left\langle \frac{n-k}{n}, \frac{n-k+1}{n} \right\rangle \cap T_n \subset \left\langle \frac{a_1}{b_1}, \frac{n-k+1}{n} \right\rangle,$$

where $a_1 = f(n-k+1, n)$ and $b_1 = \varphi(n-k+1, n)$. Therefore

$$\left\langle \frac{k-1}{n}, \frac{k}{n} \right\rangle \cap T_n = \overline{\left\langle \overline{\frac{k-1}{n}, \frac{k}{n}} \right\rangle \cap T_n} \subset \left\langle \overline{\frac{a_1}{b_1}, \frac{n-k+1}{n}} \right\rangle = \left\langle \frac{k-1}{n}, \frac{b_1 - a_1}{b_1} \right\rangle.$$

Since the fractions a_1/b_1 and $(n-k+1)/n$ are adjacent, it is easy to see that $(k-1)/n$ and $(b_1 - a_1)/b_1$ are adjacent.

c) Let Γ_n be the boundary of M_n. Then

$$\left\langle \frac{j(k,n)}{\varphi(k,n)}, \frac{k}{n} \right\rangle \cap \Gamma_n \subset T_n, \quad 0 < k < n.$$

We consider two cases.

1) $d = 1$. Suppose a_1/b_1 and a_2/b_2 are fractions adjacent to k/n. Suppose, for definiteness, $a_1/b_1 < k/n < a_2/b_2$. In the case $d = 1$, (27) can be rewritten in the form

$$\left.\begin{array}{c} b_i k - a_i n = (-1)^{i+1}, \\ a_1 + a_2 = k, \qquad b_1 + b_2 = n, \\ 0 < a_i < k, \quad 0 < b_i < n, \quad i = 1, 2. \end{array}\right\} \tag{28}$$

Suppose $n = q_1 b_1 + r_1 = q_2 b_2 + r_2$ $(0 < r_i < b_i,\ i = 1, 2)$. (It is clear from (28) that if $b_2 \neq 1$, then neither b_1 nor b_2 can be a divisor of n. We will consider the case $b_2 = 1$ later.) We have

$$\left[\frac{n-1}{b_i}\right] = q_i, \quad i = 1, 2.$$

Assume first that $b_2 < \frac{1}{2}n$ and therefore $b_1 = n - b_2 > \frac{1}{2}n$. Then $q_1 = 1$ and $q_2 > 1$, but $q_2 b_2 - b_1 = q_2 b_2 - n + b_2 = b_2 - r_2 > 0$; hence

$$\left[\frac{n-1}{b_1}\right] b_1 = b_1 < \left[\frac{n-1}{b_2}\right] b_2 = q_2 b_2.$$

Applying Theorem B to the figure M_{n-1}, we see that $\mu \in \Gamma_{n-1} \cap \langle a_1/b_1, a_2/b_2 \rangle$ satisfies the equation

$$(\mu^{b_2} - t)^{q_2} = (1 - t)^{q_2} \mu^s, \tag{29}$$

where $0 < t < 1$ and $a_2 s \equiv -1 \bmod b_2$.

Calculating by Theorem B the part of the boundary of the figure $M_{b_2 q_2}$ lying in the angle $\langle a_1/b_1, a_2/b_2 \rangle$, we see that it is also defined by (29). Therefore

$$\Gamma_{b_2 q_2} \cap \left\langle \frac{a_1}{b_1}, \frac{a_2}{b_2} \right\rangle = \Gamma_{n-1} \cap \left\langle \frac{a_1}{b_1}, \frac{a_2}{b_2} \right\rangle.$$

If we analyze analogously the case $b_1 < \frac{1}{2}n$, $b_2 = n - b_1 > \frac{1}{2}n$, we find that we always have

$$\Gamma_{n-1} \cap \left\langle \frac{a_1}{b_1}, \frac{a_2}{b_2} \right\rangle = \Gamma_m \cap \left\langle \frac{a_1}{b_1}, \frac{a_2}{b_2} \right\rangle, \tag{30}$$

where $m = [(n-1)/b]b$ and $b = \min(b_1, b_2)$.

It follows from (28) that if $b_2 = 1$, then $k = n - 1$. Applying Theorem B to this case, we see that $\mu \in \Gamma_{n-1} \cap \langle a_1/b_1, a_2/b_2 \rangle$ satisfies the equation

$$(\mu - t)^{n-1} = (1 - t)^{n-1}, \quad 0 < t < 1. \tag{31}$$

2) $d > 1$. Let a_1/b_1 be a fraction adjacent to k/n $(a_1/b_1 < k/n)$. From (27) we obtain

$$2b_1 > 2n - \frac{2n}{d} = n + n - \frac{2n}{d},$$

and since $d > 1$, it follows that $n - (2n/d) \geq 0$; hence $2b_1 > n$. Therefore $[(n-1)/b_1] = 1$. But obviously $[(n-1)/(n/d)] = d - 1$, and hence, in view of (27),

$$\left[\frac{n-1}{b_1}\right] b_1 = b_1 > \left[\frac{n-1}{(n/d)}\right] \frac{n}{d} = n - \frac{n}{d}. \tag{32}$$

It follows easily from (32), as in case 1), that

$$\Gamma_{n-1} \cap \left\langle \frac{a_1}{b_1}, \frac{k}{n} \right\rangle = \Gamma_{b_1} \cap \left\langle \frac{a_1}{b_1}, \frac{k}{n} \right\rangle. \tag{33}$$

From (30), (31), and (33) we obtain

$$\Gamma_{n-1} \cap \left\langle \frac{f(k,n)}{\varphi(k,n)}, \frac{k}{n} \right\rangle = \Gamma_{\psi(k,n)} \cap \left\langle \frac{f(k,n)}{\varphi(k,n)}, \frac{k}{n} \right\rangle,$$

where

$$\psi(k,n) = \begin{cases} \left[\frac{n-1}{b}\right] b, \; b = \min(b_1, b_2), & \text{if } d = 1 \text{ and } k \neq n-1, \\ n-1, & \text{if } d = 1 \text{ and } k = n-1, \\ b_1, & \text{if } d > 1, \end{cases}$$

and b_1, b_2 satisfy (27).

By methods analogous to those developed in [1] (see §2) we can prove that if

$$\mu \in \left\langle \frac{f(k,n)}{\varphi(k,n)}, \frac{k}{n} \right\rangle \cap \Gamma_{\psi(k,n)},$$

then there exists a convex n-gon Q such that

1) $\mu Q \subset Q$, and

2) none of the vertices of the polygon μQ lies on the contour of Q.

This shows that μ is not an extremal number for the figure M_n. So if $\lambda \in \Gamma_n \cap \langle f(k,n)/\varphi(k,n), k/n\rangle$, then $\lambda \notin \Gamma_{n-1}$.

But the boundary of the figure $M_n - M_{n-1}$ is the set-theoretic sum of the sets T_n and $\overline{T}_n$; therefore $\lambda \in T_n \cup \overline{T}_n$. However, it follows from b) that λ cannot belong to $\overline{T}_n$. Hence $\lambda \in T_n$, which proves c).

d) Items a) and c) give a complete description of the form of T_n.

The set T_n consists of n curves lying in the disc $|z| \leq 1$ and the angles $\langle f(k,n)/\varphi(k,n), k/n\rangle$, $k = 1, \ldots, n-1$. The endpoints of these arcs, taken counterclockwise, are $e^{2\pi i f(k,n)/\varphi(k,n)}$ and $e^{2\pi i k/n}$.

e) We can now proceed to a direct proof of Theorem B. Suppose a_1/b_1 and a_2/b_2 are two adjacent fractions ($a_1/b_1 < a_2/b_2$) and Γ is the boundary of M_n lying in the angle $\langle a_1/b_1, a_2/b_2\rangle$. We consider two cases.

1) Neither b_1 nor b_2 is a divisor of n. It follows from d) that Γ does not intersect T_n or $\overline{T}_n$. Therefore $\Gamma \subset M_{n-1}$ and satisfies Theorem B as a part of the boundary of M_{n-1}. We will show that Γ satisfies Theorem B as a part of the boundary of M_n. It suffices to show that $[(n-1)/b_i] = [n/b_i]$, $i = 1, 2$. But in our case these equalities are obvious.

2) At least one of b_1 and b_2 is a divisor of n.

It is easy to see that the case $n = q_2 b_2$ corresponds to case a) of Theorem B, and the case $n = q_1 b_1$ to case b). We will therefore limit ourselves to the case $n = q_2 b_2$.

It follows from c) that $\Gamma \subset T_n$; hence any $\lambda \in \Gamma$ satisfies one of the equations (12) or (15). It is easy to see, however, that (12) corresponds to I of Theorem B, and (15) to II, hence Γ satisfies Theorem B.

I would like to thank E. B. Dynkin for his generous assistance with this work.

REFERENCES

1. N. A. Dmitriev and E. B. Dynkin, *Characteristic roots of stochastic matrices*, Izv. Akad. Nauk SSSR Ser. Mat. **10** (1946), 167–184; English transl. in this volume.

Translated by G. A. KANDALL

Amer. Math. Soc. Transl.
(2) Vol. **140**, 1988

Finite Basis Property for the Identities of Representations of a Simple Three-dimensional Lie Algebra over a Field of Characteristic Zero

YU. P. RAZMYSLOV

Let E be an arbitrary Lie algebra over K, and M an associative enveloping K-algebra for E. An associative polynomial $f(x_1, \dots, x_l)$ is called an identity of the pair (E, M) if for any $v_1, \dots, v_l \in E$ one has $f|_{x_1=v_1,\dots,x_l=v_l} = 0$.

The set of all identities of a pair forms an ideal I in the free associative algebra F with free generators $x_1, x_2, \dots$, called the *ideal of weak identities*. The characteristic property of the ideal of weak identities is as follows: if $f(x_1, \dots, x_l) \in I$, then also $f(v_1, \dots, v_l) \in I$, where $v_1, \dots, v_l$ are any arbitrary elements of the Lie algebra L generated by $x_1, x_2, \dots$ in F with respect to the operation $[x, y] = xy - yx$. An ideal of identities I is called *finitely based* if there exists a finite set of polynomials in I such that the minimal ideal of identities J containing these polynomials coincides with I.

The following notation will remain fixed throughout the paper: $\mathfrak{g}$ is the three-dimensional simple (split) Lie algebra over an algebraically closed field K of characteristic zero, $\mathfrak{U}_\mathfrak{g}$ is its universal enveloping algebra, and $\mathfrak{V}_\mathfrak{g}$ is the ideal of weak identities of the pair $(\mathfrak{g}, \mathfrak{U}_\mathfrak{g})$ in the free algebra F with free generators x_1, x_2, $x_3, \dots$.

In this paper we prove the following results.

THEOREM 1. *The ideal of weak identities $\mathfrak{V}_\mathfrak{g}$ of $(\mathfrak{g}, \mathfrak{U}_\mathfrak{g})$ is finitely based.*

THEOREM 2. *Any ascending chain of ideals of weak identities each member of which contains $\mathfrak{V}_\mathfrak{g}$ stabilizes after finitely many steps.*

COROLLARY. *Any ideal of weak identities containing $\mathfrak{V}_\mathfrak{g}$ has a finite basis.*

1980 *Mathematics Subject Classification* (1985 *Revision*). Primary 16A38, 17B60.

Translation of Algebra (Collection in Honor of the Ninetieth Anniversary of the Birth of O. J. Schmidt; A. I. Kostrikin, editor). Izdat. Moskov. Gos. Univ., Moscow, 1982, pp. 139–150; MR **87c**: 17012.

§1. An auxiliary three-set algebra $(L, A, \mathscr{F}_0)$. Some identities of the pair $(\mathfrak{g}, \mathfrak{U}_\mathfrak{g})$

In this section we determine some identities of a three-set algebra of the form $(\mathfrak{g}, \mathfrak{U}_\mathfrak{g}, K)$ where $\mathfrak{g}$ is the three-dimensional simple Lie algebra over K and $\mathfrak{U}_\mathfrak{g}$ is its associative enveloping K-algebra with additional multilinear operations $\rho\colon L \to \mathfrak{U}_\mathfrak{g}$ and $(\ ,\ )\colon \mathfrak{g} \otimes_K \mathfrak{g} \to K$ satisfying the following identities:

$$\begin{gathered}\rho([x,y]) = \rho(x)\rho(y) - \rho(y)\rho(x),\\ (x,y) = (y,x), \qquad ([x,y],z) = (x,[y,z]).\end{gathered} \tag{1}$$

By definition, $(\ ,\ )$ is considered to be half of the Killing form on $\mathfrak{g}$. (We remark that $(x,y) = 2\operatorname{tr}(xy)$, where $x, y \in \mathfrak{g}$ and tr is the trace function of the two-dimensional representation of $\mathfrak{g}$.)

Let X be a set whose elements are x_1, $x_2, \ldots$.

The free algebra of the variety of the three-set algebra generated by $(\mathfrak{g}, \mathfrak{U}_\mathfrak{g}, K)$ with free generating set X, $\varnothing$, $\varnothing$ can be constructed in precisely the same way as in the case of ordinary finite-dimensional algebras.

Let e_1, e_2, e_3 be a basis of $\mathfrak{g}$ (we will assume it is an orthonormal basis with respect to $(\ ,\ )$), and $\mathscr{F}$ a commutative K-algebra with free generators of the form $x_i^{(j)}$, $j = 1, 2, 3$; $i = 1, 2, \ldots$. We set $\tilde{\mathfrak{g}} = \mathscr{F} \otimes_K \mathfrak{g}$ and $\tilde{\mathfrak{U}}_\mathfrak{g} = \mathscr{F} \otimes_K \mathfrak{U}_\mathfrak{g}$, and extend the operators ρ and $(\ ,\ )$ to $\tilde{\mathfrak{g}}$ by linearity over $\mathscr{F}$. In this case $(\tilde{\mathfrak{g}}, \tilde{\mathfrak{U}}_\mathfrak{g}, \mathscr{F})$ becomes a three-set algebra (with operation $(\ ,\ )\colon \tilde{\mathfrak{g}} \otimes \tilde{\mathfrak{g}} \to \mathscr{F}$). Let $(L, A, \mathscr{F}_0)$ denote the three-set subalgebra in $(\tilde{\mathfrak{g}}, \tilde{\mathfrak{U}}_\mathfrak{g}, \mathscr{F})$ generated by elements of the form $x_i = \sum_{j=1}^3 x_i^{(j)} e_j \in \mathfrak{g}$, and let F denote an associative subalgebra generated by x_1, $x_2, \ldots$ in $\tilde{\mathfrak{U}}_\mathfrak{g}$.

PROPOSITION 1. *The three-set algebra $(L, A, \mathscr{F}_0)$ is a relatively free three-set algebra in the variety* $\operatorname{var}(\mathfrak{g}, \mathfrak{U}_\mathfrak{g}, K)$ *with free generators X, $\varnothing$, $\varnothing$. Also, L is a relatively free Lie algebra in* $\operatorname{var}\mathfrak{g}$ *and (L, F) is a free pair in the variety of pairs of the form* $\operatorname{var}(\mathfrak{g}, \mathfrak{U}_\mathfrak{g})$ *with free generators $x_1, x_2, \ldots$.*

PROOF. This is obvious.

We pass now to the determination of some identities in $\operatorname{var}\mathfrak{g}$, $\operatorname{var}(\mathfrak{g}, \mathfrak{U}_\mathfrak{g}, K)$, and $\operatorname{var}(\mathfrak{g}, \mathfrak{U}_\mathfrak{g})$.

LEMMA 1. *Let x, y, z be elements in X. Then the following identities hold in $(\mathfrak{g}, \mathfrak{U}_\mathfrak{g}, K)$ (and in $(L, A, \mathscr{F}_0)$):*

$$[x,y,z] = x(y,z) - y(x,z), \tag{2}$$

$$v([x,y],z) = [x,y](v,z) - [x,z](y,v) + [y,z](x,v). \tag{3}$$

PROOF. It is well known that the square of a trace zero matrix of order two is a scalar matrix. Since the diagonal elements of the matrix for $(x \circ y) = xy + yx$ in the two-dimensional representation of $\mathfrak{g}$ are equal to $\frac{1}{2}\operatorname{tr}(x \circ y) = \operatorname{tr}(xy)$, we have in this representation

$$[x,y,z] = (x \circ (y \circ z)) - (y \circ (x \circ z)) = 2\operatorname{tr}(yz)x - 2\operatorname{tr}(xz)y.$$

This proves (2).

To prove (3) we apply (2) in two different ways to the commutator of weight four as follows:

$$[[x,y],v,z] = [x,y](v,z) - v([x,y],z) = [x,z](y,v) - [y,z](x,v).$$

The proof is complete.

LEMMA 2. *Let y be an arbitrary element of L. Then the following is an identity in $(\mathfrak{g}, \mathfrak{U}_\mathfrak{g}, K)$ (and in $(L, A, \mathscr{F}_0)$):*

$$[x_i, x_j] \cdot (y,y) = [x_i, y, y, x_j] + [y, x_j, x_i, y]. \tag{4}$$

Moreover, for any multilinear polynomial $g(x_1, \dots, x_l) \in F$ $(F \subset A)$ which is skew-symmetric over x_i and x_j we have

$$g \cdot (y,y) = \tfrac{1}{2}(g|_{x_i=[x_i,y,y]} + g|_{x_j=[x_j,y,y]} + g|_{x_i=[x_i,x_j,y];x_j=y}). \tag{5}$$

PROOF. Applying (2) to commutators of the form $[x_i, y, y]$ and $[y, x_j, x_i]$, we find that the right-hand side of (4) takes the form

$$[x_i,x_j](y,y) - [y,x_j](x_i,y) + [y,y](x_j,x_i) - [x_i,y](y,x_j) = [x_i,x_j](y,y).$$

This proves (4).

Applying (2) to $[x_i, y, y]$, $[x_j, y, y]$, and $[x_i, x_j, y]$, we see that the right-hand side of (5) has the form

$$\tfrac{1}{2}(g(y,y) - g|_{\substack{x_i=y\\x_j=x_j}}(x_i,y) + g(y,y) - g|_{\substack{x_i=x_i\\x_j=y}}(x_j,y) \\ + g|_{\substack{x_i=x_i\\x_j=y}}(x_j,y) - g|_{\substack{x_i=x_j\\x_j=y}}(x_i,y)).$$

In this expression the fourth and the fifth terms differ only in sign; and, since g is skew-symmetric in x_i and x_j, the sum of the second and sixth terms is zero. The proof is complete.

REMARK 1. For any u, v, x, $y \in L$, let $[u,v] * (x,y)$ denote the Lie element of the form

$$[u,x,y,v] + [u,y,x,v] + [[x,v],u,y] + [y,v,u,x].$$

This element emerges as a complete linearization of the right-hand side in (4). Let $\langle v_1, v_2, v_3\rangle$ denote

$$\sum_{s \in S_3} \varepsilon(s) v_{s(1)} v_{s(2)} v_{s(3)}$$

and let $z * \langle \operatorname{ad} v_1, \operatorname{ad} v_2, \operatorname{ad} v_3\rangle$ denote a Lie element of the form

$$\sum_{s \in S_3} \varepsilon(s) [z, v_{s(1)}, v_{s(2)}, v_{s(3)}].$$

LEMMA 3. *The following are weak identities in $(\mathfrak{g}, \mathfrak{U}_\mathfrak{g})$ (and in (L, F)):*

$$[x,y,z] \cdot [u,v] = x([u,v] * (y,z)) - y([u,v] * (x,z)), \tag{6}$$

$$(z * \langle \operatorname{ad} x_1, \operatorname{ad} x_2, \operatorname{ad} x_3\rangle) y = z(y * \langle \operatorname{ad} x_1, \operatorname{ad} x_2, \operatorname{ad} x_3\rangle), \tag{7}$$

$$\langle x_1, x_2, x_3\rangle y = y \langle x_1, x_2, x_3\rangle. \tag{8}$$

PROOF. Since F is in A, we may use (2), (4), and Remark 1 for the proof of (6). Now (8) is verified by direct substitution of the elements of the orthonormal basis e_1, e_2, e_3 in place of x_1, x_2, x_3. Then $\langle e_1, e_2, e_3 \rangle$ becomes a multiple of the Casimir element $e_1^2 + e_2^2 + e_3^2$. Finally, (7) follows from Schur's lemma and (8) applied to the adjoint representation of $\mathfrak{g}$. The proof is complete.

LEMMA 4. $(\mathfrak{g}, \mathfrak{U}_{\mathfrak{g}})$ *(and* (L, F)*) satisfies the Capelli identities of order four.*

PROOF. We need to prove that an arbitrary multilinear polynomial which is skew-symmetric with respect to a set of four variables vanishes after its variables are replaced by elements of $\mathfrak{g}$. This is, however, obvious since $\mathfrak{g}$ is three-dimensional.

LEMMA 5. *All Lie identities of* $\mathfrak{g}$ *follow from a single identity of degree five of the form*

$$[(z * \langle \operatorname{ad} x_1, \operatorname{ad} x_2, \operatorname{ad} x_3 \rangle), y] = [z, (y * \langle \operatorname{ad} x_1, \operatorname{ad} x_2, \operatorname{ad} x_3 \rangle)]. \tag{9}$$

PROOF. It follows from the weak identity (7) that (9) holds in $\mathfrak{g}$. Now it has been shown in [1] that all identities of $\mathfrak{g}$ follow from those of degree five. It has been shown in [2] and [3] that the space of all multilinear Lie polynomials of degree five as a module over the permutation group S_5 decomposes as a direct sum of five pairwise nonisomorphic irreducible representations such that the identities of our algebra constitute the direct sum of two of them and one of these latter corresponds to a Young diagram of the form

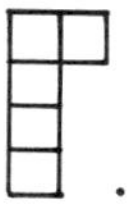

This diagram corresponds to a Lie standard identity which is a consequence of (9). It remains to remark that this Lie standard identity holds in the Lie algebra of derivations of the polynomial ring in one variable, while (9) is not an identity of this Lie algebra. The proof is complete.

§2. Proof of Theorem 2

This section discloses the true reason for the relative Specht property of $\operatorname{var}(\mathfrak{g}, \mathfrak{U}_{\mathfrak{g}})$. Roughly speaking, this is due to the representability of subvarieties of $\operatorname{var}(\mathfrak{g}, \mathfrak{U}_{\mathfrak{g}})$ in subvarieties of $\operatorname{var}(\mathfrak{g}, \mathfrak{U}_{\mathfrak{g}}, K)$, and the relative Specht property of the latter is obvious.

PROOF. Let (L, F) denote the free algebra in $\operatorname{var}(\mathfrak{g}, \mathfrak{U}_{\mathfrak{g}})$, and let

$$I_1 \subseteq I_2 \subseteq I_3 \subseteq \cdots \tag{10}$$

be an ascending chain of ideals of weak identities in F. Let J denote the ideal of weak identities determined by the commutativity identity. To prove that (10) stabilizes it obviously suffices to prove the same for the chain whose terms are

$J \cap I_i$. Indeed, F/J is commutative and any set of identities in F/J is equivalent to an identity of the form $x^m = 0$ for an appropriate m. Accordingly, it will be assumed in the sequel that $I_i \subset J$ for all i.

LEMMA 6. *Let $(L, A, \mathscr{F}_0)$ and (L, F) be as in Proposition 1. Then, for any ideal of weak identities I in J, the linear subspace $I + I \cdot \mathscr{F}_0$ is contained in I. In other words, any ideal of weak identities contained in J is an $\mathscr{F}_0$-module.*

PROOF. Since the characteristic of the base field K is zero, it is sufficient to show that for any multilinear element $g(x_1, \dots, x_l) \in I \subset J$ and for any element of the form $f = (x_{l+1}, x_{l+2})$ we have $g * f \in I$.

By Lemma 2, for any transposition τ_{ij}, $i, j \in \{I, \dots, l\}$, and any permutation $s \in S_l$ we have

$$\{(1 - \tau_{ij})s(g)\} * f = h_{s,i,j}, \tag{11}$$

where $h_{s,i,j} \in I$. Since the two-sided ideal T of the group ring $K[S_l]$ of the symmetric group S_l generated by the elements of the form $1 - \tau_{ij}$, $i, j = 1, \dots, l$, is the linear span of the elements $s - \delta$, where $s, \delta \in S_l$, it follows from (11) that, for any permutation $s \in S_l$,

$$\{g - s(g)\} * f = h_s, \tag{12}$$

where $h_s \in I$. Summing (12) over all $s \in S_l$, we obtain

$$n!g * f - \sum_{s \in S_l} s(g) * f = \sum h_s \in I.$$

But for any multilinear polynomial $g \in J$ we have $\sum_{s \in S_l} s(g) = 0$, proving the lemma.

Since $(\mathfrak{g}, \mathfrak{U}_\mathfrak{g})$ satisfies a system of weak Capelli identities it is well known [4] that Theorem 2 will be proved if we establish the stabilization property for the chain (10) in the case where (L, F) (respectively, $(L, A, \mathscr{F}_0)$) has finitely many generators $x_1, \dots, x_k$.

In this case we consider J as a left module over the algebra $D \subset \operatorname{End}_K J$ generated by multiplications on the elements of $\mathscr{F}_0$ and by left multiplications on the elements of F.

LEMMA 7. *J is a finitely generated D-module.*

PROOF. Any element in J can be represented as a K-linear combination of the products av, where v is a left-normed commutator of degree at least two and a is a word in x_i $(i = 1, \dots, k)$. It follows then by (2) that, as a D-module, J is generated by the left-normed commutators of degrees 2 and 3 in $x_1, x_2, \dots$. The proof is complete.

LEMMA 8. *D is a left (right) Noetherian K-algebra.*

PROOF. The K-algebra D is in fact an $\mathscr{F}_0$-algebra. It follows from (2) that $\mathscr{F}_0$ is generated by (u, v), where u and v are commutators of degree at most three. This means that $\mathscr{F}_0$ is a finitely generated commutative K-algebra. Let

L denote the Lie subalgebra of D generated by left multiplication by elements of L. Then $\tilde{L} = L + L \cdot \mathscr{F}_0$ is a finitely generated $\mathscr{F}_0$-module. It follows from (2) and (6) that this module is generated by commutators of degree at most two.

Our claim now follows from the fact that the enveloping algebra of a Lie algebra which is a finitely generated module over a commutative Noetherian ring Φ is itself left (right) Φ-Noetherian. The proof is complete.

It follows from Lemmas 6–8 that the D-module J is Noetherian and that the ascending chain (10) of ideals of weak identities is an ascending chain of D-submodules in J. Therefore, starting from some N, for any $j > 0$ we have $I_{N+j} = I_N$, proving Theorem 2.

§3. Proof of Theorem 1

In this section we give an explicit form of weak identities which enable us to successfully apply the argument in §2. To prove Theorem 1 it will be necessary, as a matter of fact, to repeat the proof of Theorem 2.

Let (L, F_1) denote the relatively free pair with free generators $x_1, x_2, \dots$, determined by (6) and (9). It follows from Lemmas 3 and 5 that L is a free algebra in var $\mathfrak{g}$.

REMARK 2. Let $(L, A, \mathscr{F}_0)$ be as in Proposition 1. As we know, the derived algebra L' of L is an $\mathscr{F}_0$-algebra (this follows from (9) in Lemma 5). Moreover, this action of $\mathscr{F}_0$ commutes with each $\operatorname{ad} v$, $v \in L$, restricted to L'.

Our first aim is to prove that the ideal J of weak identities determined by the commutativity in F_1 is also an $\mathscr{F}_0$-module.

The proof splits into several steps.

LEMMA 9. *If $u, v \in L'$, $y \in L$, $a \in F_1$, and $f = (y, y) \in \mathscr{F}_0$, then the following is a weak identity in F_1:*

$$(u * f)av = ua(v * f). \tag{13}$$

PROOF. It is sufficient to consider the case when $u = [x_1, x_2]$, $v = [z_1, z_2]$, and a is some word in free generators. If a is a nonempty word then $a = a_1 x_i$, where a_1 is a shorter word. In this case we have

$$(u * f)a_1 x_i v = (u * f)a_1[x_i, v] + (u * f)a_1 v x_i. \tag{14}$$

Using induction and Remark 2, we find that the right-hand side of this equation has the form

$$ua_1[x_i, v * f] + ua_1(v * f)x_i = ua_1 x_i(v * f).$$

It follows that we may restrict ourselves to the case where a in (13) is empty. In this case we have

$$\begin{aligned}([x_1, x_2] * f)[z_1, z_2] &= \{[x_1, y, y, x_2] + [y, x_2, x_1, y]\}[z_1, z_2] \\ &= [[x_1, y, y] \cdot [z_1, z_2], x_2] - [x_1, y, y] \cdot [z_1, z_2, x_2] \\ &\quad + [[y, x_2, x_1] \cdot [z_1, z_2], y] - [y, x_2, x_1][z_1, z_2, y].\end{aligned}$$

By (6) and Remark 2 the right-hand side of this equation takes the form

$$[x_1, x_2]([z_1, z_2] * (y, y)) - [y, x_2]([z_1, z_2] * (x_1, y)) \\ + [y, y]([z_1, z_2] * (x_2, x_1)) - [x_2, y]([z_1, z_2] * (y, x_1)).$$

The third term in this expression is equal to zero, and the second and fourth differ only in sign. The proof is complete.

LEMMA 10. *Let a multilinear polynomial $g(x_1, \ldots, x_l) \in F_1$ have the form*

$$g = \sum_k a_k v_k,$$

where $v_k \in L'$ and $a_k \in F_1$, and let g be skew-symmetric in the two variables x_i and x_j. Then

$$\tfrac{1}{2}\{g|_{x_i=[x_i,y,y]} + g|_{x_j=[x_j,y,y]} + g|_{x_i=[x_i,x_j,y];x_j=y}\} = \sum_k a_k(v_k * (y, y)). \quad (14)$$

PROOF. It suffices to consider the case $g = (1 - \tau_{ij})av$, where v is a left-normed commutator of degree at least two, a a word in $x_1, x_2, \ldots$, and $\tau_{ij} \in S_1$ a transposition. If x_i and x_j occur only in v, then (14) follows from Lemma 2 and Remark 2. If x_i and x_j occur only in a, then $(1 - \tau_{ij})a$ can be represented in the form $\sum_k a'_k v'_k$, where $v'_k \in L'$. Since a is of degree less than l, the inductive argument enables us to represent the left-hand side of (14) in the form

$$\sum_k a'_k(v'_k * f)v,$$

and by Lemma 9 this is equal to

$$\sum_k a'_k v'_k(v * (y, y)) = (1 - \tau_{ij})\{a\{v * (y, y)\}\}.$$

It remains to consider the case where

$$g = x_i a[x_j, \ldots] - x_j a[x_i, \ldots].$$

Exactly in the same way as in the proof of Lemma 9, using induction and Remark 2, we may assume that a is an empty word. And, in this particular case, the proof of (14) can be carried out precisely in the same way as the proof of (5), except for the use of (6) and Remark 2 in place of (2) in decomposing the commutators $[x_i, y, y]$, $[x_j, y, y]$, and $[x_i, x_j, y]$. The proof is complete.

LEMMA 11. *Let g be an arbitrary polynomial in the ideal J of weak identities determined by commutativity in F_1 and I_g the least ideal of weak identities of F_1 containing g. Let g be represented in the form $\sum_k a_k v_k$, where $v_k \in L'$. Then for any $f \in \mathscr{F}_0$ the polynomial $\sum_k a_k(v_k * f)$ is an element of I_g.*

The proof is essentially the same as for Lemma 6. Since the characteristic of the base field K is zero, it is sufficient to prove our claim in the case where $g(x_1, \ldots, x_l)$ is an arbitrary multilinear polynomial in J, $f = (x_{l+1}, x_{l+2})$, and

g has the form $\sum_k a_k v_k$. By Lemma 10, for any transposition τ_{ij} and $s \in S_l$ we have

$$(1-\tau_{ij})s\left(\sum_k a_k(v_k * f)\right) = h_{s,i,j},$$

where $h_{s,i,j} \in I_g$. It follows that for any permutation $s \in S_l$

$$(1-s)\left(\sum_k a_k(v_k * f)\right) = h_s, \tag{15}$$

where $h_s \in I_g$. Summing (15) over all $s \in S_l$, we get

$$l!\sum_k a_k(v_k * f) - \sum_{s\in S_l} s(a_k(v_k * f)) = \sum_{s\in S_l} h_s.$$

The second term in the left-hand side of this equation is equal to zero in F_1, since the Lie polynomial of the form

$$\sum_{s\in S_m}([x_{s(1)},\dots,x_{s(m)}] * f),$$

with the summation taken over all permutations $s \in S_m$, is an identity of a three-dimensional simple Lie algebra $\mathfrak{g}$ (see Remark 2). The proof is complete.

PROPOSITION 2. *Let a relatively free pair (L, F_1) be determined by (6) and (9), and let $\mathscr{F}_0$ be as in Proposition 1. Then the ideal of weak identities determined by commutativity is an $\mathscr{F}_0$-module with respect to the action given by*

$$\left(\sum_k a_k v_k\right)\cdot f = \sum_k a_k(v_k * f), \tag{16}$$

where $a_k \in F_1$, $v_k \in L'$, and $f \in \mathscr{F}_0$. Moreover, any ideal of weak identities I contained in J is an $\mathscr{F}_0$-submodule.

PROOF. We have to verify that the action of $\mathscr{F}_0$ on J given by (16) does not depend on the choice of representation for $g \in J$ in the form $\sum_k a_k v_k$. But this is a consequence of Lemma 11, proving our claim.

THEOREM 3. *Any set of polynomials in k variables $x_1,\dots,x_k$ in F_1 determines an ideal of weak identities with finite basis.*

PROOF. Repeat the arguments in Lemmas 7 and 8.

COROLLARY. *If a variety of pairs satisfies weak identities (6) and (9) and a system of Capelli identities of some order, then it is finitely based.*

PROOF. Capelli identities of order m and Theorem 3 ensure the stabilization property for the chains of ideals of weak identities in the free pair given by Capelli identities of order m. It remains to note that the system of Capelli identities of arbitrary order m is finitely based [5]. The proof is complete.

PROOF OF THEOREM 1. By Lemmas 3–5, $\operatorname{var}(\mathfrak{g}, \mathfrak{U}_\mathfrak{g})$ satisfies the system of Capelli identities of order four, and also (6) and (8). Hence this variety is finitely based.

REMARK 3. It is not hard to show that all Capelli identities of order four follow from (6)–(9) together with the standard identity of order four

$$\sum_{s\in S_4} \varepsilon(s)x_{s(1)}x_{s(2)}x_{s(3)}x_{s(4)} = 0, \qquad (17)$$

satisfied in $(\mathfrak{g}, \mathfrak{U}_{\mathfrak{g}})$. Therefore, the set of identities (6)–(9) and (17) determine a variety of pairs which is Specht. The proof of this fact is left to the reader. (Hint: use Proposition 2.)

REFERENCES

1. Yu. P. Razmyslov, *The existence of a finite basis for the identities of the matrix algebra of order two over a field of characteristic zero*, Algebra i Logika **12** (1973), 83–113; English transl. in Algebra and Logic **12** (1973).

2. A. A. Klyachko, *Lie elements in a tensor algebra*, Sibirsk. Mat. Zh. **15** (1974), 1296–1304; English transl. in Siberian Math. J. **15** (1974).

3. V. S. Drenski, *Identities in matrix Lie algebras*, Trudy Sem. Petrovsk. Vyp. 6 (1981), 47–55; English transl. in J. Soviet Math. **6** (1986), no. 3.

4. Yu. P. Razmyslov, *The Jacobson radical in PI-algebras*, Algebra i Logika **13** (1974), 337–360; English transl. in Algebra and Logic **13** (1974).

5. ——, *Algebras satisfying Capelli identities*, Izv. Akad. Nauk SSSR Ser. Mat. **45** (1981), 143–166; English transl. in Math. USSR Izv. **18** (1982).

Translated by YU. BAKHTURIN

Amer. Math. Soc. Transl.
(2) Vol. **140**, 1988

On the Number of Nontrivial Solutions of Homogeneous Equations in Words with Three Unknowns

L. G. KISELEVA

A coefficient-free equation in words with three unknowns is an equality of a pair of distinct nonempty words

$$F(x, y, z) = G(x, y, z) \tag{1}$$

in the alphabet of unknowns $\{x, y, z\}$.

The equation (1) is said to be *homogeneous* if each unknown appears the same number of times in the left and right sides; otherwise the equation is said to be *nonhomogeneous.*

A triple of words $\langle x_0, y_0, z_0\rangle$ in some alphabet is a *solution* if their substitution into (1) results in a graph-theoretical identity.

Khmelevskiĭ [3] showed that all the solutions of (1) can be enumerated in a finite number of formulas. For instance, the general solution of the equation $xy = zx$ is described by the set $\langle (uv)^a u, (vu)^b, (uv)^b\rangle$, where u and v are arbitrary words, and a and b are nonnegative integers, with $b > 0$.

Two solutions, $\langle x_0, y_0, z_0\rangle$ and $\langle x_0', y_0', z_0'\rangle$, are said to be *isomorphic* if they satisfy the same set of equations of the form (1).

A solution $\langle x_0, y_0, z_0\rangle$ is said to be *trivial* (a *T-solution*) if the set of words appearing in it has the form $\{\alpha^i, \alpha^j, \beta\}$ or $\{\alpha, \beta, \varphi(\alpha, \beta)\}$, where $\varphi(\alpha, \beta)$ is either the empty word or some product of the words α and β. The other solutions are called *nontrivial* ($\overline{T}$*-solutions*).

It is known [2] that for the equation $xy = zx$ all $\overline{T}$-solutions are isomorphic to the solution $\langle a, ba, ab\rangle$.

Let $\overline{N}(F, G)$ be the number of pairwise nonisomorphic $\overline{T}$-solutions of (1). For the previous example, we then have $\overline{N}(xy, zx) = 1$.

1980 *Mathematics Subject Classification* (1985 *Revision*). Primary 20M05.
Translation of Combinatorial-Algebraic Methods in Applied Mathematics (A. A. Markov, editor), Gor′kov. Gos. Univ., Gorki, 1981, pp. 73–76; MR **85i**: 20038.

It is shown in [2] that for every nonhomogeneous equation (1),

$$\overline{N}(F, G) = 0 \tag{2}$$

or

$$\overline{N}(F, G) = 1. \tag{3}$$

It is shown in the same article that for homogeneous equations other structures of pairwise nonisomorphic solutions are possible:

$$\overline{N}(F, G) = \infty. \tag{4}$$

For instance, (4) is satisfied for the equations $xzxy = yxzx$ and $xyz = zxy$.

For arbitrary equations in words with two unknowns,

$$F(x, y) = G(x, y), \tag{5}$$

where $F(x, y)$ and $G(x, y)$ are distinct words in the alphabet of unknowns $\{x, y\}$, the following theorem establishes the form of the solutions and shows that the problem of finding all solutions of (5) reduces to solving a certain diophantine equation.

THEOREM 1 [1], [4]. *If $\langle x_0, y_0\rangle$ is a solution of* (5), *then $x_0 = \alpha^i$ and $y_0 = \alpha^j$ for some word α and natural numbers i and j.*

COROLLARY. *If $x_0 = \alpha^i$, $y_0 = \alpha^j$, $z_0 = \beta$ constitute a solution of* (1), *then there exist a word γ and numbers k, l, m such that $x_0 = \gamma^k$, $y_0 = \gamma^l$, and $z_0 = \gamma^m$.*

It is shown in [5] that

$$\overline{N}(z^n, x^k y^l) = 0 \tag{6}$$

if $n, k, l \geq 2$.

This assertion is a special case of a theorem proved in [1].

THEOREM 2. *$\overline{N}(z^n, \varphi(x, y)) = 1$ if and only if $\varphi(x, y) = (x^i y x^j)^k$ or $\varphi(x, y) = (y^i x y^i)^k$, with $i + j > 0$ and n not a divisor of k.*

One can show that (4) holds for all homogeneous equations of the form (1) with $|F| + |G| = 6$ (where $|F|$ is the word length of F), whereas for $|F| + |G| = 8$, 10, either (2) or (4) is satisfied.

In this note we give an example of a homogeneous equation for which (3) is satisfied.

Consider the equation

$$x^2 z y^2 z = y^2 z^2 x^2. \tag{7}$$

THEOREM 3. $\overline{N}(x^2 z y^2 z, y^2 z^2 x^2) = 1$.

To prove this, it will be enough to show that all the solutions of (7) are solutions of a nonhomogeneous equation, and that (7) admits $\overline{T}$-solutions (by virtue of properties (2) and (3) of nonhomogeneous equations). It is immediately verified that $x_0 = a^2b$, $y_0 = a$, $z_0 = ba^2b$ is a $\overline{T}$-solution of (7). We will show that all solutions of (7) are solutions of $x^2 = y^2z$.

Right multiplication of both sides of (7) by z yields the equivalent equation

$$x^2zy^2z^2 = y^2z^2x^2z. \tag{8}$$

We will denote x^2z by u and y^2z^2 by v. In this new notation, (8) then takes on the form

$$uv = vu. \tag{9}$$

Hence, if $\langle x_0, y_0, z_0\rangle$ is a solution of (7), then $x_0^2z_0 = u_0$ and $y_0^2z_0^2 = v_0$, i.e., $\langle u_0, v_0\rangle$ is a solution of (8), and, by Theorem 1, $u_0 = \alpha^i$ and $v_0 = \alpha^j$ for some word α and numbers i, j. Then $\langle \alpha, y_0, z_0\rangle$ is a $\overline{T}$-solution of $y^2z^2 = w^j$ (otherwise the corollary to Theorem 1 would not be satisfied), i.e., $j = 1$, by Theorem 2. We will show that every solution of (7) is a solution of the equation $x^2z = (y^2z^2)^i$, or of the equivalent equation $x^2 = (y^2z^2)^{i-1}y^2z$. Since (7) has a $\overline{T}$-solution, the last equation must also have a $\overline{T}$-solution, and this is impossible unless $i = 1$ (by Theorem 2).

So, the proof is complete.

REMARK. The peculiarities of homogeneous equations are not typical of all classes of equations. Almost all equations are nonhomogeneous in the following sense. The *length* of equation (1) is defined to be $|F| + |G|$. Let $l(n)$ denote the number of equations of length n and of the form (1), and $l_0(n)$ the number of such equations that are homogeneous. Then,

$$l_0(n)/l(n) \longrightarrow 0 \quad \text{as } n \to \infty. \tag{10}$$

If n is odd, then $l_0(n) = 0$.

We will show that for $n = 2k$, the following holds:

$$l(n) = (n-1)3^n - 3^{n/2}, \tag{11}$$

$$l_0(n) < 3^n. \tag{12}$$

Let us consider the word

$$x_{i_1}x_{i_2}\cdots x_{i_n} \tag{13}$$

in the alphabet $\{x, y, z\}$. There are 3^n such words, and each one can be broken into a pair of nonempty words

$$x_{i_1}x_{i_2}\cdots x_{i_r}, \qquad x_{i_{r+1}}x_{i_{r+2}}\cdots x_{i_n} \tag{14}$$

in $n-1$ ways. There are $3^n(n-1)$ possible pairs of words of the form (14), among which $3^{n/2}$ are identical.

Every pair of distinct words of the form (14) gives rise to an equation (1) of length n. Therefore, $l(n)$ satisfies (11).

The number of all pairs of words of the form (14) with $r = k$ equals 3^n, and some of these words define homogeneous equations (1) of length n. This implies (12).

Suppose that (11) and (12) are satisfied. Then

$$\frac{l_0(n)}{l(n)} < \frac{3^n}{(n-1)3^n - 3^{n/2}} = \frac{1}{(n-1) - 3^{-n/2}} \leq \frac{1}{n-2} \to 0 \quad \text{as } n \to \infty$$

since $3^{-n/2} \leq 1$. This proves (10).

In conclusion, the author would like to express his gratitude to Al. A. Markov for his help with this article.

REFERENCES

1. L. G. Budkina and Al. A. Markov, *On F-semigroups with 3 generators*, Mat. Zametki **14** (1973), 267–277; English transl. in Math. Notes **14** (1973).

2. L. G. Kiseleva, *An algebraic study of the simplest codes and coefficientless equations in words*, Mat. Sb. **108 (150)** (1979), 529–550; English transl. in Math. USSR Sb. **36** (1980).

3. Yu. I. Hmelevskiĭ, *Equations in free semigroups*, Trudy Mat. Inst. Steklov. **107** (1971); English transl., Proc. Steklov Inst. Math. **107** (1971) (1976).

4. E. K. Blum, *A note on free subsemigroups with two generators*, Bull. Amer. Math. Soc. **71** (1965), 678–679.

5. R. C. Lyndon and M. B. Shützenberger, *The equation* $a^M = b^N c^P$ *in a free group*, Michigan Math. J. **9** (1962), 289–298.

Translated by D. GILDENHUYS

Amer. Math. Soc. Transl.
(2) Vol. **140**, 1988

On the Representation of Monoids by Multivalued Mappings

L. A. SKORNYAKOV

Let REL be the category of nonvoid sets with binary correspondences as morphisms. In accordance with the general concept described in [3], a pair (A, f), where A is a nonvoid set and f a homomorphism of a monoid R into $H_{\mathrm{REL}}(A, A)$, is called a REL-representation of R. This is in complete agreement with the definition in [5] of a representation of the monoid by multivalued mappings. The category of REL-representations of R is then defined in the natural fashion; it will be denoted by REL-Act-R. Two monoids R and S are called REL-equivalent if the categories REL-Act-R and REL-Act-S are equivalent. A natural question arises: does REL-equivalence of R and S imply their Morita equivalence [6], [7]? We recall that the converse implication follows from results of Polin ([2], Theorem 1). We recall also that, as shown by Banaschewski and Knauer (see Proposition 1), Morita equivalence of two monoids is characterized by semigroup properties of the monoids. The results of the present paper can therefore be regarded as a first step toward the solution of the above problem. Namely, we prove:

THEOREM. *If the monoids R and S are* REL*-equivalent, then their global monoids 2^R and 2^S are Morita equivalent.*

The question of the converse implication remains open. We note that it is closely related to the following problem: are two monoids isomorphic if their global monoids are? A negative answer was given by Mogilyanskaya [1] in the case of semigroups. But her example fails to carry over to monoids.

We shall use the following notation:

Act-R is the category of right R-polygons.

1980 *Mathematics Subject Classification* (1985 *Revision*). Primary 18B10; Secondary 20M30, 20M50.

Translation of Algebra (Collection in Honor of the Ninetieth Anniversary of the Birth of O. J. Schmidt; A. I. Kostrikin, editor), Izdat. Moskov. Gos. Univ., Moscow, 1982, pp. 151–157; MR **86f:** 20084.

SET-equivalence of monoids means their Morita equivalence.

1_A is the identity mapping of a set A onto itself.

$\delta_A = \{(x,x)|x \in A\}$.

If $\varphi \subseteq A \times B$ and $X \subseteq A$, then

$$X\varphi = \{y|y \in B,\ \exists x \in X,\ (x,y) \in \mu\}.$$

If $\varphi \subseteq A \times B$ and $\psi \subseteq B \times C$, then

$$\varphi\psi = \{(a,c)|a \in A,\ c \in C,\ \exists b \in B\ ((a,b) \in \varphi)\ \&\ ((b,c) \in \psi)\}.$$

If (A,f) is an object in REL-Act-R, then for $af(\lambda)$, where $a \in A$ and $\lambda \in R$, we shall often write simply $a\lambda$; and by $\mathrm{End}(A,f)$ we shall mean the set $H_{\text{REL-Act-}R}((A,f),(A,f))$ with multiplication $\varphi\psi$ defined by the equality

$$\varphi\psi = \{(a',a'')|a',a'' \in A,\ \exists x \in A\ ((a',x) \in \psi)\ \&\ ((x,a'') \in \varphi)\}.$$

For convenience of reference we state here:

PROPOSITION 1 [6, p. 61, PROPOSITION 4; 7, p. 367, THEOREM 6.1]. *SET-equivalence of two monoids R and S is equivalent to existence of elements ε, λ, $\mu \in R$ such that $\varepsilon = \varepsilon^2$, $\lambda\varepsilon = \lambda$, $\lambda\mu = 1$, and $S \cong \varepsilon R\varepsilon$. The equivalence of the categories Act-R and Act-S is effected by the functor $T = H_{\text{Act-}R}(\varepsilon R, -)$.*

PROPOSITION 2. *Let R be a monoid; (A,f) and (B,g) two REL-representations of R; $\varphi\colon (A,f) \to (B,g)$ a morphism in the category REL-Act-R; and (R,e) the REL-representation such that $\xi e(\lambda) = \xi\lambda$ for any $\lambda, \xi \in R$. Then the following hold.*

(a) *If $X \in 2^A$ and $\Lambda \in 2^R$, then $(Xf(\Lambda))\varphi = (X\varphi)g(\Lambda)$.*

(b) *For every $X \in 2^A$ there exists a unique morphism $\chi\colon (R,e) \to (A,f)$ such that $1\chi = X$.*

(c) *The monoids $\mathrm{End}(R,e)$ and 2^R are isomorphic.*

(d) *If φ is a monomorphism, $a', a'' \in A$ and $a'\varphi = a''\varphi$, then $a' = a''$.*

(e) *If φ is an epimorphism, then the property*

$$\forall b \in B \quad \exists a \in A \quad (a\varphi = b) \tag{$*$}$$

is necessary and sufficient for φ to be a strong epimorphism.

(f) The representation (R,e) is projective relative to strong epimorphisms.

(g) The coproduct of the representations (A_i, f_i), $i \in I$, is the representation (A,f), where A is the free union of the sets A; and $af(\lambda) = af_i(\lambda)$, $a \in A_i$, $\lambda \in R$.

(h) For every representation (A,f), there exists a coproduct (F,h) of a set of copies of the representation (R,e) and a strong epimorphism $\pi\colon (F,h) \to (A,f)$.

(i) If (A,f) is a coproduct of representations (A_i, f_i), $i \in I$, and $\pi\colon (A,f) \to (R,e)$ is a strong epimorphism, then the restriction of π to some A_i is also a strong epimorphism.

(j) If $\pi\colon (R,e) \to (A,f)$ is a strong epimorphism and $X = 1\pi$, then

$$\forall a \in A \quad \exists \lambda \in R \quad (a = Xf(\lambda)).$$

PROOF. (a) This follows from the following equivalences:

$$b \in (Xf(\Lambda))\varphi \Leftrightarrow \exists x \in X\ \exists \lambda \in \Lambda\ (b \in xf(\lambda)\varphi = x\varphi g(\lambda)) \\ \Leftrightarrow b \in (X\varphi)g(\lambda).$$

(b) It suffices to put $\xi\chi = Xf(\xi)$ for each $\xi \in R$ and to observe that

$$1\chi = Xf(1) = X\delta_A = X,$$

$$\xi(e(\lambda)\chi) = (\xi\lambda)\chi = Xf(\xi\lambda) = (Xf(\xi))f(\lambda) = (\xi\chi)f(\lambda)$$

for any $\lambda, \xi \in R$.

(c) Define the mapping $\Gamma\colon \mathrm{End}(R, e) \to 2^R$ by putting $\Gamma(\varphi) = \varphi(1)$. That Γ is one-to-one is easily shown by using (b). Furthermore, if $\varphi, \psi \in \mathrm{End}(R, e)$, we have by (a) that

$$\Gamma(\varphi\psi) = \varphi\psi(1) = \varphi(1\psi(1)) = \varphi(1)\psi(1) = \Gamma(\varphi)\Gamma(\psi).$$

(d) Suppose $a'\varphi = a''\varphi$ but $a' \neq a''$. By (b), there exist morphisms $\psi\colon (R, e) \to (A, f)$ and $\chi\colon (R, e) \to (A, f)$ such that $1\psi = a'$ and $1\chi = a''$. Then $\psi \neq \chi$ but $\psi\varphi = \chi\varphi$, contradicting the assumption that φ is a monomorphism.

(e) If $(*)$ holds, consider the commutative diagram

$$\begin{array}{ccc} (A, f) & \xrightarrow{\varphi} & (B, g) \\ \psi \downarrow & & \downarrow \chi \\ (C, h) & \xrightarrow{\theta} & (D, k), \end{array}$$

where θ is a monomorphism. For each $b \in B$ put $b\rho = a\psi$, where $a\varphi = b$. If $a'\varphi = b$, then

$$a'\psi\theta = a'\varphi\chi = a\varphi\chi = a\psi\theta.$$

Hence $a'\psi = a\psi$, by (d); i.e., the correspondence ρ is well defined. Clearly ρ is a morphism in the category REL-Act-R, and $\varphi\rho = \psi$. In addition, $\varphi\rho\theta = \psi\theta = \varphi\chi$, so that $\rho\theta = \chi$. Thus, φ is a strong epimorphism ([4], pp. 39, 40). Conversely,(1) if φ is strong, define on the set A an equivalence $\equiv$ by putting $x \equiv y$ if $x\varphi = y\varphi$. Let $\overline{A} = A/\equiv$, let ψ be the natural mapping of A onto $\overline{A}$, and let

$$(a\psi)\overline{f}(\lambda) = (af(\lambda))\psi$$

for any $a \in A$ and $\lambda \in R$. Then $(a\psi)(\overline{f}(1)) = a\psi$, and if $\lambda, \mu \in R$, then

$$(a\psi)\overline{f}(\lambda\mu) = (af(\lambda\mu))\psi = ((af(\lambda))f(\mu))\psi \\ = ((af(\lambda))\psi)\overline{f}(\mu) = (a\psi)(\overline{f}(\lambda)\overline{f}(\mu)).$$

(1) *Translator's note.* The converse argument appears to me to be incorrect (and I don't see how to correct it). The operation of R on $\overline{A}$ needn't be well defined; and if it is, the mapping θ needn't be a monomorphism, even if $R = \{1\}$. [Editor's note. The author responds that the relations under consideration are everywhere defined, but the proof of Proposition 2(e) cannot be rescued. He is also unable to give another characterization of morphisms $\rho \in A \times B$ of the category REL-Act-R, where $(a, b'), (a, b'') \in \rho$ implies $b' = b''$ and for each $b \in B$ there exists $a \in A$ such that $(a, b) \in \rho$, in the language of category theory. Therefore the theorem is not proved.]

Consequently, $(\overline{A}, \overline{f})$ is a REL-representation of R, and ψ a morphism in the category REL-Act-R. Putting $\overline{a}\theta = a\varphi$, we obtain a commutative diagram

$$\begin{array}{ccc} (A, f) & \xrightarrow{\varphi} & (B, g) \\ \psi \downarrow & & \downarrow \delta_B \\ (\overline{A}, \overline{f}) & \xrightarrow{\theta} & (B, g), \end{array}$$

where θ is a monomorphism. Since φ is strong, there exists a morphism $\rho: (B, g) \to (\overline{A}, \overline{f})$ giving two commutative triangles. Consequently, we have for any $b \in B$ that $b\rho = X\psi$ for some $X \in 2^A$; so that $b = b\delta_B = b\rho\theta = X\psi\theta = X\varphi\delta_B = X\varphi$. Thus, $b = a\varphi$ for all $a \in X \subseteq A$.

(f) Given the diagram

$$\begin{array}{ccc} & & (R, e) \\ & & \downarrow \psi \\ (A, f) & \xrightarrow{\varphi} & (B, g), \end{array}$$

where φ is a strong epimorphism, we have by (e) that $1\psi = X\varphi$, where $X \in 2^A$. By (b), there exists a morphism $\chi: (R, e) \to (A, f)$ such that $1\chi = X$. From (b) also follows the equality $\chi\varphi = \psi$.

(g) Clearly (A, f) is a REL-representation; and if $\varphi_i: (A_i, f_i) \to (B, g)$, $i \in I$, then putting $a\varphi = a\varphi_i$ for $a \in A_i$ gives a morphism $\varphi: (A, f) \to (B, g)$ whose restriction to A_i coincides with φ_i for every $i \in I$.

(h) Let (F, h) be the coproduct of the representations (R_a, e_a), each equal to (R, e). By (b), there exist morphisms $\varphi_a: (R_a, e_a) \to (A, f)$ such that $1\varphi_a = a$. Extending them to a morphism $\varphi: (F, h) \to (A, f)$, we see, using (e), that φ is a strong epimorphism.

(i) By (e), $1 = a\pi$ for some $a \in A$. But then, by (g), $a \in A_i$ for some $i \in I$; and this has the desired properties.

(j) If $a \in A$, then, by (e), $a = \lambda\pi$ for some $\lambda \in R$. Hence

$$a = (1\lambda)\pi = (1\pi)f(\lambda) = Xf(\lambda).$$

PROPOSITION 3. *Let (A, f) be a* REL*-representation of a monoid R. Put $\Gamma(A, f) = 2^A$ and $X\Lambda = Xf(\Lambda)$ for any $X \in 2^A$ and $\Lambda \in 2^R$. Then 2^A becomes a right 2^R-polygon. If $\varphi: (A, f) \to (B, g)$ is a morphism in the category* REL-Act-*R, then putting $X\Gamma(\varphi) = X\varphi$ for every $X \in 2^A$ gives a morphism $\Gamma(\varphi): 2^A \to 2^B$ in the category* Act-2^R. *Furthermore, every morphism $\overline{\varphi}: \Gamma(A, f) \to \Gamma(B, g)$ in* Act-2^R *is of the form $\overline{\varphi} = \Gamma(\varphi)$ for some morphism $\varphi: (A, f) \to (B, g)$ in* REL-Act-*R.*

PROOF. The first assertion is verified by a simple calculation. For the second, Proposition 2(a) gives the equality

$$(X\Lambda)\Gamma(\varphi) = (Xf(\Lambda))\varphi = (X\varphi)g(\Lambda) = (X\Gamma(\varphi))\Lambda,$$

showing that $\Gamma(\varphi)$ is a morphism in Act-2^R. Finally, if $\overline{\varphi}$ is as indicated, put

$$\varphi = \{(x, y) | x \in A,\ y \in B,\ y \in x\overline{\varphi}\}.$$

Then for any $x \in A$ and $\lambda \in R$ we have the equivalences

$$\begin{aligned}(y \in x(f(\lambda)\varphi)) &\Leftrightarrow (\exists u \in A((u \in xf(\lambda)) \;\&\; (y \in u\varphi))) \\ &\Leftrightarrow (\exists u \in A((u \in x\lambda) \;\&\; (y \in u\overline{\varphi}))) \\ &\Leftrightarrow (y \in (x\lambda)\overline{\varphi} = (x\overline{\varphi})\lambda) \\ &\Leftrightarrow (\exists v \in B((v \in x\overline{\varphi}) \;\&\; (y \in v\lambda))) \\ &\Leftrightarrow (\exists v \in B((v \in x\varphi) \;\&\; (y \in vg(\lambda)))) \\ &\Leftrightarrow (y \in x(\varphi g(\lambda))),\end{aligned}$$

showing that $f(\lambda)\varphi = \varphi g(\lambda)$. Consequently, φ is a morphism in REL-Act-R; and clearly $\Gamma(\varphi) = \overline{\varphi}$.

PROOF OF THE THEOREM. Suppose the categories REL-Act-R and REL-Act-S are equivalent, and let (A, f) be the object in REL-Act-R that corresponds to the object (S, e) under this equivalence. From Proposition 2(h) and 2(i) it follows that there exists a strong epimorphism $\psi\colon (R, e) \to (A, f)$. Proposition 2(f) then guarantees the existence of a morphism $\varphi\colon (A, f) \to (R, e)$ such that $\varphi\psi = \delta_A$. Similarly, there exists a diagram

$$(R, e) \xrightarrow{\chi} (A, f) \xrightarrow{\rho} (R, e),$$

where ρ is a strong epimorphism and $\chi\rho = \delta_R$.

Put $E = 1\psi\varphi$, $\Lambda = 1\psi\rho$, and $M = 1\chi\varphi$. Clearly $E, \Lambda, M \in 2^R$. Furthermore, using Proposition 2(a), we find that

$$\begin{aligned}E^2 = (1\psi\varphi)(e(E)) &= (1e(E))\psi\varphi = E\psi\varphi \\ &= 1\psi\varphi\psi\varphi = 1\psi\varphi = E, &\quad (*)\\ \Lambda E = (1\psi\rho)(e(E)) &= (1e(E))\psi\rho = E\psi\rho \\ &= 1\psi\varphi\psi\rho = 1\psi\rho = \Lambda, &\quad (**)\\ \Lambda M = (1\psi\rho)(e(M)) &= (1e(M))\psi\rho = M\psi\rho \\ &= 1\chi\varphi\psi\rho = 1\chi\rho = 1\delta_R = 1. &\quad (***)\end{aligned}$$

Now put $X = 1\psi$ and, using Proposition 3, consider the right 2^R-polygon $\Gamma(A, f) = 2^A$. By Proposition 2(j), $2^A = X2^R$. Consider the mapping $\theta\colon 2^A \to E2^R$ given by $(XN)\theta = EN$ for all $N \in 2^R$. The mapping is well defined. Indeed, if $XN = XN'$, where $N, N' \in 2^R$, then by Proposition 3, $Xf(N) = Xf(N')$; so, by Proposition 2(a),

$$\begin{aligned}EN &= (1\psi\varphi)N = (X\varphi)N = (Xf(N))\varphi \\ &= (Xf(N'))\varphi = (X\varphi)N' = (1\psi\varphi)N' = EN'.\end{aligned}$$

On the other hand, if $EN = EN'$, then

$$\begin{aligned}XN &= Xf(N) = (1\psi\varphi\psi)f(N) = (E\psi)f(N) \\ &= (EN)\psi = (EN')\psi = (E\psi)f(N') = (1\psi\varphi\psi)f(N') = Xf(N') = XN'.\end{aligned}$$

Thus, the right 2^R-polygons 2^A and $E2^R$ are isomorphic. By Propositions 2(c) and 3, and [7], p. 366, Lemma 5.2, we have the monoid isomorphisms

$$\begin{aligned}2^S &\cong \mathrm{End}_{\text{REL-Act-}S}(S, e) \cong \mathrm{End}_{\text{REL-Act-}R}(A, f) \\ &\cong \mathrm{End}_{\text{Act-}2^R}\, 2^A \cong \mathrm{End}_{\text{Act-}2^R}(E2^R) \cong E2^RE.\end{aligned}$$

From this and equalities $(*)$, $(**)$, and $(* * *)$, Proposition 1 gives the SET-equivalence of the monoids 2^R and 2^S.

BIBLIOGRAPHY

1. E. M. Mogiljanskaja [Mogilyanskaya], *Non-isomorphic semigroups with isomorphic semigroups of subsets*, Semigroup Forum **6** (1973), 330–333.

2. S. V. Polin, *Morita-equivalent categories*, Vestnik Moskov. Univ. Ser. I Mat. Meh. **29** (1974), no. 2, 41–45; English transl. in Moscow Univ. Math. Bull. **29** (1974).

3. L. A. Skornyakov, *A general view of representations of monoids*, Algebra i Logika **20** (1981), 571–574; English transl. in Algebra and Logic **20** (1981).

4. M. Sh. Tsalenko and E. G. Shul′geĭfer, *Foundations of category theory*, "Nauka", Moscow, 1974. (Russian)

5. B. M. Shaĭn [Schein], *Representation of semigroups by means of binary relations*, Mat. Sb. (N.S.) **60 (102)** (1963), 293–303. (Russian)

6. Bernhard Banaschewski, *Functors into categories of M-sets*, Abh. Math. Sem. Univ. Hamburg **38** (1972), 49–64.

7. Ulrich Knauer, *Projectivity of acts and Morita equivalence of monoids*, Semigroup Forum **3** (1971/72), 359–370.

Translated by J. A. ZILBER

Amer. Math. Soc. Transl.
(2) Vol. **140**, 1988

Representation of Functions in Convex Domains by Generalized Exponential Series

A. F. LEONT'EV

Dedicated to Academician B. Szökefalvi-Nagy on his seventieth birthday

Abstract. Let D be a finite convex domain. It is known (see [1]) that any function $F(z)$ analytic in D can be represented by an exponential series

$$F(z) = \sum_{n=1}^{\infty} A_n e^{\lambda_n z}, \quad z \in D.$$

The exponents λ_n depend only on the domain D and have finite upper density, and the series converges absolutely in D, and uniformly on compact subsets.

Here we give a class A of functions $f(z)$ of exponential type having the property that any function $\Phi(z)$ analytic in the convex domain D ($0 \in D$) can be represented in D as a series

$$\Phi(z) = \sum_{n=1}^{\infty} B_n f(\lambda_n z), \quad z \in D,$$

which converges uniformly on compact subsets of D.

1. Definition of the class A. By definition, a function $f(z) = \sum_{n=0}^{\infty} a_n z^n/n!$ belongs to the class A if $a_n \neq 0$ ($n \geq 0$) and the functions

$$\eta(t) = \sum_{n=0}^{\infty} \frac{a_n}{t^{n+1}}, \qquad \eta_1(t) = \sum_{n=0}^{\infty} \frac{1}{a_n t^{n+1}}$$

are regular outside the interval $[0, 1]$ of the real axis.

The class A contains the function e^z; in this case

$$\eta(t) = \eta_1(t) = \sum_{n=0}^{\infty} \frac{1}{t^{n+1}} = \frac{1}{t-1}.$$

1980 *Mathematics Subject Classification* (1985 *Revision*). Primary 30B50, 30B60.
Translation of Acta Sci. Math. (Szeged) **45** (1983), 305–315; MR **84i**:30004.

We have that

$$\overline{\lim_{n\to\infty}} \sqrt[n]{|a_n|} \le 1, \qquad \overline{\lim_{n\to\infty}} \sqrt[n]{\left|\frac{1}{a_n}\right|} \le 1,$$

which implies that $\lim_{n\to\infty} \sqrt[n]{|a_n|} = 1$. Consequently, $f(z)$ is an entire function of exponential type equal to 1.

We remark that $\eta(t)$ is the function associated with $f(z)$ in the Borel sense.

2. A sufficient test for membership of a function in the class A.

THEOREM 1. *Suppose that $a_n = \varphi(n)$ and $1/a_n = \varphi_1(n)$, $n \ge N$, where $\varphi(z)$ and $\varphi_1(z)$ are functions analytic in the half-plane* $\operatorname{Re} z \ge N$ *and*

$$|\varphi(z)| < e^{\varepsilon|z|}, \ |\varphi_1(z)| < e^{\varepsilon|z|}, \ \operatorname{Re} z \ge N, \ |z| > r_0(\varepsilon), \ \forall\varepsilon > 0. \tag{1}$$

Then $f(z) \in A$.

PROOF. Let $0 < \varphi_0 < \pi/2$, and let $p \ge N$ be a noninteger. Let q be the point where the ray $\arg z = \varphi_0$ intersects the line $\operatorname{Re} z = p$. Denote by Γ the contour formed from the rays $[q, \infty e^{i\varphi_0})$ and $[\overline{q}, \infty e^{-i\varphi_0})$ and the segment $[q, \overline{q}]$. Define

$$\psi(t) = \frac{1}{2i}\int_\Gamma \frac{\varphi(z)e^{-zt}\,dz}{\sin \pi z}. \tag{2}$$

By (1),

$$\left|\frac{\varphi(z)}{\sin \pi z}\right| < A(\varepsilon)e^{(-\pi\sin\varphi_0+\varepsilon)|z|} \quad \forall\varepsilon > 0 \tag{3}$$

on the rays in the definition of Γ. This implies that the integral (2) converges and is an analytic function in the angle $B(\varphi_0)$ with one side passing through the point πi at an angle $\pi/2 - \varphi_0$ to the real axis and the other side symmetric to the first with respect to the real axis.

An estimate of the form (3) remains valid on the arcs $|z| = r_m = m\pi + \frac{\pi}{2}$ $(m = 1, 2, \dots)$, $|\arg z| \le \varphi_0$. Therefore, if Γ_m is the closed contour formed from the part of Γ lying in the disk $|z| \le r_m$ and the arcs $|z| = r_m$ and $|\arg z| \le \varphi_0$, then for t in the angle $|\arg t| < \frac{\pi}{2} - \varphi_0$ we have

$$\psi(t) = \lim_{m\to\infty} \frac{1}{2i}\int_{\Gamma_m} \frac{\varphi(z)e^{-zt}\,dz}{\sin \pi z} = \sum_{n>p} (-1)^n \varphi(n) e^{-nt}.$$

Observe now that

$$\psi(t) = \frac{1}{2i}\int_{\operatorname{Re} z=p} \frac{\varphi(z)e^{-zt}\,dz}{\sin \pi z}. \tag{4}$$

This integral converges in the strip $|\operatorname{Im} t| < \pi$. Hence, $\psi(t)$ is a function analytic in this strip and in the half-plane $\operatorname{Re} t > 0$. This implies that the function

$$\psi_1(t) = \sum_{n=0}^{\infty} (-1)^n a_n e^{-nt}$$

is also analytic in the strip $|\operatorname{Im} t| < \pi$ and in the half-plane $\operatorname{Re} t > 0$. In view of this, the function $\eta(t)$ is analytic outside the segment $[0, 1]$.

It can also be proved that $\eta_1(t)$ is regular outside $[0,1]$. Therefore, $f(z) \in A$.

REMARK. It follows from the representation (4) that

$$|\psi(t)| < C|e^{-pt}|, \qquad |\operatorname{Im} t| < \pi - \delta, \quad \delta > 0. \tag{5}$$

This implies that outside the angle $|\arg t| < \delta$ (as $t \to 0$)

$$|\eta(t)| < \frac{C}{|t|^{p+1}}, \qquad |\eta_1(t)| < \frac{C}{|t|^{p+1}} \quad (C = C(\delta)). \tag{6}$$

3. A converse of Theorem 1.

THEOREM 2. *Suppose that the functions $\eta(t)$ and $\eta_1(t)$ are regular outside the segment $[0,1]$ and satisfy condition (6) outside each angle $|\arg t| < \delta$. Then there are functions $\varphi(z)$ and $\varphi_1(z)$ analytic for $\operatorname{Re} z \geq N$ and satisfying condition (1) such that $a_n = \varphi(n)$ and $\frac{1}{a_n} = \varphi_1(n)$, $n \geq N$.*

PROOF. Let

$$\psi(t) = \sum_{n>p} (-1)^{n-1} a_n e^{-nt}. \tag{7}$$

According to the conditions of the theorem, this function is regular in the half-plane $\operatorname{Re} t > 0$ and in the strip $|\operatorname{Im} t| < \pi$, and it has the estimate (5) in the smaller strip $|\operatorname{Im} t| < \pi - \delta$ (as $t \to -\infty$). As $t \to +\infty$ in the strip $|\operatorname{Im} t| < \pi - \delta$ we have

$$|\psi(t)| < C|e^{-p_1 t}|, \qquad p_1 > p. \tag{8}$$

We consider the integral

$$\varphi(z) = \frac{\sin \pi z}{\pi} \int_{-\infty+ic}^{\infty+ic} \psi(t) e^{zt}\, dt, \qquad -\pi < c < \pi. \tag{9}$$

In view of (5) and (8) it converges for z in the strip $p < \operatorname{Re} z < p_1$, and the value does not depend on the parameter c. We have that

$$\varphi(z) = \frac{\sin \pi z}{\pi} \int_{-\infty}^{0} \psi(t) e^{tz}\, dt + \frac{\sin \pi z}{\pi} \int_{0}^{\infty} \psi(t) e^{tz}\, dt = \varphi_1(z) + \varphi_2(z).$$

The function $\varphi_1(z)$ is regular in the half-plane $\operatorname{Re} z > p$, and $\varphi_1(n) = 0$ for $n > p$. The function $\varphi_2(z)$ is regular in the half-plane $\operatorname{Re} z < p_1$.

Take $0 < \varphi_0 < \frac{\pi}{2}$. We have that

$$\begin{aligned} \varphi_2(z) &= \frac{\sin \pi z}{\pi} \int_{0}^{\varepsilon e^{i\varphi_0}} \psi(t) e^{tz}\, dt + \frac{\sin \pi z}{\pi} \int_{\varepsilon e^{i\varphi_0}}^{\infty e^{i\varphi_0}} \psi(t) e^{tz}\, dt \\ &= \frac{\sin \pi z}{\pi} \varphi_3(z) + \varphi_4(z). \end{aligned}$$

The function $\varphi_3(z)$ is an entire function of exponential type at most ε. Let us study $\varphi_4(z)$. When $\arg z = \varphi$ satisfies the condition $\frac{\pi}{2} < \varphi + \varphi_0 < \frac{3\pi}{2}$, we get that

$$\begin{aligned} \varphi_4(z) &= \frac{\sin \pi z}{\pi} \int_{\varepsilon e^{i\varphi_0}}^{\infty e^{i\varphi_0}} \left(\sum_{n>p} (-1)^{n-1} a_n e^{-nt} \right) e^{tz}\, dt \\ &= -\frac{\sin \pi z}{\pi} e^{\varepsilon_0 z} \sum_{n>p} (-1)^{n-1} \frac{a_n e^{-\varepsilon_0 n}}{z-n}, \qquad \varepsilon_0 = \varepsilon e^{i\varphi_0}. \end{aligned}$$

The right-hand side is an entire function of exponential type, and it has value equal to a_n at the point $z = n$. Moreover,

$$|\varphi_4(x)| < e^{2\varepsilon x}, \qquad x > x_0(\varepsilon).$$

In summary, the function $\varphi(z)$ is regular and of exponential type in the half-plane $\operatorname{Re} z > p$, and $\varphi(n) = a_n$ $(n > p)$ and

$$|\varphi(x)| < e^{\varepsilon x}, \qquad x > x_1(\varepsilon), \qquad \forall \varepsilon > 0. \tag{10}$$

Let $z = p_2 + iy$, $p < p_2 < p_1$. By (9),

$$|\varphi(z)| < M|\sin \pi z| e^{-cy}, \qquad -\pi < c < \pi.$$

Choosing the quantity c close to π for $y > 0$ and close to $-\pi$ for $y < 0$, we get that

$$|\varphi(z)| < B(\varepsilon) e^{\varepsilon |y|}, \qquad z = p_2 + iy, \qquad \forall \varepsilon > 0.$$

From this and (10) we obtain

$$|\varphi(z)| < e^{\varepsilon |z|}, \qquad \operatorname{Re} z \geq p_2, \qquad |z| > r_0(\varepsilon), \qquad \forall \varepsilon > 0.$$

The function $\varphi(z)$ has all the necessary properties. The existence of the needed function $\varphi_1(z)$ has also been proved.

4. Two examples. In the first example

$$f(z) = e^z + ae^{qz}, \quad 0 < q < 1.$$

Here

$$a_n = 1 + aq^n, \qquad \frac{1}{a_n} = \frac{1}{1 + aq^n}, \qquad \varphi(z) = 1 + ae^{z \ln q}, \qquad \varphi_1(z) = \frac{1}{1 + ae^{z \ln q}}$$
$$(\ln q < 0).$$

We see that the functions $\varphi(z)$ and $\varphi_1(z)$ have the necessary properties, and thus $f(z) \in A$.

In the second example,

$$f(z) = \sum_{n=0}^{\infty} \frac{a_n}{n!} z^n, \qquad a_n = 1 + (-1)^n e^{-n^2} \quad (n \geq 0).$$

We have that

$$\eta(t) = \frac{1}{t-1} + \sum_{n=0}^{\infty} \frac{(-1)^n e^{-n^2}}{t^{n+1}}, \qquad \eta_1(t) = \frac{1}{t-1} - \sum_{n=0}^{\infty} \frac{(-1)^n e^{-n^2}}{(1 + (-1)^n e^{-n^2}) t^{n+1}}.$$

The series converge for all $t \neq 0$. Therefore, $\eta(t)$ and $\eta_1(t)$ have only the two singularities $t = 0$ and $t = 1$; hence $f(z) \in A$. We remark that for $t < 0$ the function

$$\eta(t) = \frac{1}{t-1} + \frac{1}{t} \sum_{n=0}^{\infty} \left(-\frac{1}{t}\right)^n e^{-n^2}$$

tends to ∞ more rapidly than any power of $1/t$ as $t \to 0$. In view of this there are no functions $\varphi(z)$ and $\varphi_2(z)$ with the properties indicated above.

5. The transformation M. By definition, the transformation M carries the function $F(z) = \sum_{n=0}^{\infty} A_n z^n$ into

$$\Phi(z) = M(F) = \sum_{n=0}^{\infty} B_n z^n, \qquad B_n = a_n A_n \quad (n \geq 0).$$

THEOREM 3. *Suppose that $f(z) = \sum_{n=0}^{\infty} a_n z^n/n!$ belongs to the class A. If the function $F(z)$ is regular in a domain E which is star-like with respect to the origin of coordinates, then the function $\Phi(z) = M(F)$ is also regular in E, and if K is a compact subset of E, and C is a closed contour lying in E, encircling K, and star-like with respect to the origin, then*

$$|\Phi(z)| < N \max_{t \in C} |F(t)|, \quad z \in K, \tag{11}$$

where the constant N does not depend on $F(z)$.

PROOF. We consider the function

$$\psi(z) = \frac{1}{2\pi i} \int_C \eta\left(\frac{t}{z}\right) F(t) \frac{dt}{z}, \tag{12}$$

where C is a closed contour star-like with respect to the origin and lying in E, and z lies inside C. Then t/z cannot fall on the segment $[0, 1]$; therefore, $\psi(z)$ is an analytic function inside C. But C is an arbitrary contour in E, and hence $\psi(z)$ is regular in E.

Let $z \in K$ and $t \in C$. Then $|\eta(\frac{t}{z})| \leq N_1$, and it follows from (12) that

$$|z\psi(z)| \leq N_1 l \max_{t \in C} |F(t)|$$

(l is the length of the contour C). It can be assumed that K contains some disk $|z| \leq \delta$. Then

$$|\psi(z)| \leq \frac{N_1 l}{\delta} \max_{t \in C} |F(t)|, \quad z \in K,\ |z| \geq \delta,$$

$$|\psi(z)| \leq \frac{N_1 l}{\delta} \max_{t \in C} |F(t)|, \quad |z| \leq \delta$$

(the second inequality is valid by the maximum modulus principle). Thus,

$$|\psi(z)| \leq N \max_{t \in C} |F(t)|, \quad z \in K,\ N = N_1 l/\delta.$$

It remains to prove that $\psi(z) = \Phi(z)$. Suppose that $|z|$ is sufficiently small. Then

$$\eta\left(\frac{t}{z}\right) = \sum_{n=0}^{\infty} \frac{a_n z^{n+1}}{t^{n+1}}, \quad t \in C,$$

$$\psi(z) = \sum_{n=0}^{\infty} a_n \left(\frac{1}{2\pi i} \int_\Gamma \frac{F(t)\, dt}{t^{n+1}}\right) z^n = \sum_{n=0}^{\infty} a_n A_n z^n = \Phi(z).$$

It can be established similarly that if $\Phi(z)$ is regular in E, then the function $F(z) = M^{-1}(\Phi)$ is also regular in E, and

$$|F(z)| < N_1 \max_{t \in C} |\Phi(t)|, \quad z \in K. \tag{13}$$

We note the following properties of the transformation M.

1. $M(F_1 + F_2) = M(F_1) + M(F_2)$.
2. $M(cF) = cM(F)$.
3. If the functions $F_n(z)$ $(n \geq 1)$ are regular in a domain E (star-like with respect to the origin) and $\{F_n(z)\}$ converges to $F(z)$ uniformly on compact subsets of E, then $M(F_n) \to M(F)$ uniformly on compact subsets of E.
4. If the functions $\Phi_n(z)$ are regular in E and $\{\Phi_n(z)\}$ converges to $\Phi(z)$ uniformly on compact subsets of E, then $M^{-1}(\Phi_n) \to M^{-1}(\Phi)$ uniformly on compact subsets of E.
5. $M(e^{\lambda z}) = f(\lambda z)$.

Corollaries 3 and 4 follow on the basis of (11) and (13).

6. Series expansions. As indicated at the beginning of the paper, each function $\Phi(z)$ analytic in a finite convex domain D can be represented in the form

$$F(z) = \sum_{n=1}^{\infty} A_n e^{\lambda_n z}, \quad z \in D \tag{14}$$

(the exponents depend only on the domain D, and the convergence is uniform on compact subsets of D).

THEOREM 4. *Suppose that D is a finite convex domain, $0 \in D$, and $f(x) \in A$. Then each function $\Phi(z)$ analytic in D can be represented in the form*

$$\Phi(z) = \sum_{n=1}^{\infty} A_n f(\lambda_n z) \tag{15}$$

(*uniform convergence on compact subsets of D*).

PROOF. Let $F(z) = M^{-1}(\Phi)$. The function $F(z)$ is representable by the series (14). Then

$$\Phi(z) = M(F) = M\left(\sum_{n=1}^{\infty} A_n e^{\lambda_n z}\right) = \sum_{n=1}^{\infty} A_n M(e^{\lambda_n z}) = \sum_{n=1}^{\infty} A_n f(\lambda_n z), \quad z \in D.$$

7. Coefficient formulas. Let $A(\overline{D})$ be the class of functions analytic on the closure $\overline{D}$. In the case when $F(z) \in A(\overline{D})$ there are formulas for determining the coefficients A_n in (14). Namely, let $L(\lambda)$ be an entire function of exponential type and of completely regular growth with growth indicator $h(\varphi) = K(-\varphi)$ ($K(\varphi)$ is the support function of the domain $\overline{D}$) and with simple zeros $\lambda_n = |\lambda_n| e^{i\varphi_n}$ $(n \geq 1)$, and assume that

$$|L'(\lambda_n)| > e^{[h(\varphi_n) - \varepsilon]|\lambda_n|}, \quad n > n_0(\varepsilon), \ \forall \varepsilon > 0.$$

Then in (14) we can take the exponents to be the zeros λ_n of this function $L(\lambda)$, and the coefficients to be the quantities

$$A_n = \frac{1}{2\pi i} \int_C \psi_n(t) F(t)\, dt \quad (n \geq 1),$$

where the $\psi_n(t)$ are the functions associated in the Borel sense with the functions

$$\frac{L(\lambda)}{(\lambda-\lambda_n)L'(\lambda_n)}$$

(they are regular outside $\overline{D}$), and C is a closed contour encircling $\overline{D}$ on which and inside which $F(t)$ is an analytic function.

We have that

$$A_n = \frac{1}{2\pi i}\int_C \psi_n(t)M^{-1}(\Phi)\,dt = \frac{1}{2\pi i}\int_C \psi_n(t)\left\{\frac{1}{2\pi i}\int_{C_1}\eta_1\left(\frac{u}{t}\right)\Phi(u)\frac{du}{t}\right\}dt.$$

Here C_1 is a closed convex contour encircling C on which and inside which $\Phi(u)$ is an analytic function. Changing the order of integration, we get that

$$A_n = \frac{1}{2\pi i}\int_{C_1}\gamma_n(u)\Phi(u)\,du \quad (n \geq 1), \tag{16}$$

where

$$\gamma_n(u) = \frac{1}{2\pi i}\int_C \psi_n(t)\eta_1\left(\frac{u}{t}\right)\frac{dt}{t}.$$

The functions $\gamma_n(u)$ are regular outside $\overline{D}$, and $\gamma_n(\infty) = 0$. We note that

$$\frac{1}{2\pi i}\int_{C_1}\gamma_n(u)f(\lambda_m u)\,du = \frac{1}{2\pi i}\int_C \psi_n(t)e^{\lambda_m t}\,dt = \delta_{nm}.$$

Thus, $\{\gamma_n(u)\}$ is a system biorthogonal to the system $\{f(\lambda_n z)\}$. With its help the coefficients in (15) can be determined by (16) (under the condition that $\Phi(z) \in A(\overline{D})$).

8. An example of a function $f(z)$ for which Theorem 4 does not hold. Let $f(z) = e^z + e^{qz}$. Then

$$\eta(t) = \frac{1}{t-1} + \frac{1}{t-q}.$$

If $q \notin [0,1]$, then $f(z)$ does not belong to the class A (for $q \in [0,1]$ it belongs to A). We show that in this case there exist convex domains D ($0 \in D$) in which it is not possible to represent an arbitrary analytic function $\Phi(z)$ on D by a series (15).

We have that

$$f(z) = 2e^{\mu z}\cos\lambda z, \qquad \mu = \frac{1+q}{2}, \qquad \lambda = \frac{1-q}{2i}.$$

LEMMA 1. *Outside the neighborhoods* $|z \mp \frac{k\pi - \pi/2}{\lambda}| < \frac{1}{k^\alpha}$ ($k \geq 1$, $\alpha > 0$) *and* $|z| > 1$

$$|f(z)| > \frac{B}{|z|^\alpha}\max(|e^z|, |e^{qz}|), \quad B > 0, \tag{17}$$

where B is a constant.

PROOF. Outside the indicated neighborhoods

$$|\cos\lambda z| > \frac{B}{|z|^\alpha}\exp\{|\mathrm{Im}(\lambda z)|\}, \quad B > 0.$$

Let z lie outside these neighborhoods. In the case $\operatorname{Im}(\lambda z) \geq 0$,

$$|f(z)| > \frac{B}{|z|^\alpha} \exp\{\operatorname{Re}(\mu z) + \operatorname{Im}(\lambda z)\}.$$

But ($\mu = i\lambda + q$)

$$\operatorname{Re}(\mu z) + \operatorname{Im}(\lambda z) = \operatorname{Re}(qz) + \operatorname{Re}(i\lambda z) + \operatorname{Im}(\lambda z) = \operatorname{Re}(qz).$$

Moreover ($\mu = 1 - i\lambda$),

$$\operatorname{Re}(\mu z) + \operatorname{Im}(\lambda z) \geq \operatorname{Re}(\mu z) = \operatorname{Re} z - \operatorname{Re}(i\lambda z) = \operatorname{Re} z + \operatorname{Im}(\lambda z) \geq \operatorname{Re} z.$$

Inequality (17) follows from this.

In the case $\operatorname{Im}(\lambda z) < 0$,

$$|f(z)| > \frac{B}{|z|^\alpha} \exp\{\operatorname{Re}(\mu z) - \operatorname{Im}(\lambda z)\}.$$

In this case ($\mu = 1 - i\lambda$)

$$\operatorname{Re}(\mu z) - \operatorname{Im}(\lambda z) = \operatorname{Re} z - \operatorname{Re}(i\lambda z) - \operatorname{Im}(\lambda z) = \operatorname{Re} z$$

and, moreover ($\mu = i\lambda + q$)

$$\operatorname{Re}(\mu z) - \operatorname{Im}(\lambda z) \geq \operatorname{Re}(\mu z) = \operatorname{Re}(qz) - \operatorname{Im}(\lambda z) \geq \operatorname{Re}(qz).$$

We again obtain (17).

LEMMA 2. *Suppose that* $\overline{\lim}_{n\to\infty} \frac{n}{|\lambda_n|} < \infty$ *and the series* $\sum_{n=1}^{\infty} A_n f(\lambda_n z)$ *converges in the neighborhood* $|z - z_0| < \delta$ $(z_0 \neq 0)$. *Then the series*

$$\sum_{n=1}^{\infty} A_n e^{\lambda_n z}, \qquad \sum_{n=1}^{\infty} A_n e^{q\lambda_n z} \tag{18}$$

converge in some neighborhood $|z - z_0| < \delta_0 < \delta$.

PROOF. Outside the disks

$$C_k^{(n)}\colon \left| z \mp \frac{k\pi - \pi/2}{\lambda_n} \right| < \frac{1}{|\lambda_n| k^\alpha} \quad (k \geq 1)$$

we have the estimate

$$|f(\lambda_n z)| > \frac{B}{|\lambda_n z|^\alpha} \max(|e^{\lambda_n z}|, |e^{q\lambda_n z}|), \tag{19}$$

by Lemma 1.

Let $0 < r < |z_0| - \delta < |z_0| + \delta < R$. The disks $C_k^{(n)}$ which intersect the neighborhood $K\colon |z - z_0| < \delta$ are such that, for $n > n_0$,

$$r < \frac{k\pi}{|\lambda_n|} < R.$$

Thus, $\frac{r|\lambda_n|}{\pi} < k < \frac{R|\lambda_n|}{\pi}$, and hence the sum of the diameters of the $C_k^{(n)}$ intersecting K does not exceed the quantity

$$\frac{R|\lambda_n|}{\pi} \cdot \frac{2}{|\lambda_n| \left(\frac{r|\lambda_n|}{\pi}\right)^\alpha} = \frac{a}{|\lambda_n|^\alpha}.$$

The sum of the diameters of the indicated disks as n varies from N to ∞ does not exceed

$$\alpha_N \equiv a \sum_{n=N}^{\infty} \frac{1}{|\lambda_n|^\alpha} \to 0, \qquad N \to \infty, \quad \alpha > 1.$$

Suppose that $\alpha_N < \delta$. Then there exists a circle $|z - z_0| = \delta_1$, $0 < \delta_1 < \delta$, which does not intersect the disks under consideration, and hence (19) holds on it for $n \geq N$. On this circle take points γ_1, γ_2, and γ_3 forming a triangle E containing z_0 in its interior. According to the condition,

$$|A_n f(\lambda_n \gamma_j)| \leq M \quad (n \geq N;\ j = 1, 2, 3).$$

By (19), this gives us that

$$|A_n e^{\lambda_n \gamma_j}| < M_1 |\lambda_n|^\alpha, \qquad |A_n e^{q\lambda_n \gamma_j}| < M_1 |\lambda_n|^\alpha \quad (n \geq N).$$

We take into account next that in view of the inequality $a^x b^{1-x} < a + b$, $a > 0$, $b > 0$, $0 \leq x \leq 1$, the relation

$$|e^{\lambda_n z}| \leq |e^{\lambda_n z_1}| + |e^{\lambda_n z_2}|$$

holds for points z in the segment $[z_1, z_2]$ $(z = \beta z_1 + (1 - \beta) z_2,\ 0 \leq \beta \leq 1)$. Therefore,

$$|A_n e^{\lambda_n z}| \leq 3M_1 |\lambda_n|^\alpha, \qquad |A_n e^{q\lambda_n z}| \leq 3M_1 |\lambda_n|^\alpha \tag{20}$$

on the boundary of the triangle E, and hence inside it. Inside E we take a circle $|z - z_0| = \rho$ and write for the points of this circle

$$|A_n e^{\lambda_n z_0}| = |A_n e^{\lambda_n z}| |e^{-\lambda_n (z - z_0)}|.$$

Let $|z - z_0| = \rho$ and $\arg(z - z_0) = -\arg \lambda_n$. Then, by (20),

$$|A_n e^{\lambda_n z_0}| \leq 3M_1 |\lambda_n|^\alpha e^{-\rho|\lambda_n|}, \quad n \geq N.$$

Hence, the first of the series (18) converges at z_0. We see similarly that the second series also converges at z_0. Accordingly, the convergence of $\sum_{n=1}^{\infty} A_n f(\lambda_n z)$ in the disk $|z - z_0| < \delta$ implies the convergence of the series (18) at the center of the disk. But then, in view of this, (18) converge in some neighborhood $|z - z_0| < \delta_0$.

LEMMA 3. *Suppose that D is the domain of convergence of the series* (15), *and G and G_q are the domains of convergence of the series* (18). *Then $D = G \cap G_q$.*

The proof is obvious in view of Lemma 2.

Let D be the rectangle $|\mathrm{Re}\, z| < \varepsilon$, $-h_1 < \mathrm{Im}\, z < h_2$, $h_2 > h_1 > 0$. We take the function $\Phi(z) = 1/(z - \varepsilon)$. Assume that the expansion (15) is valid. If $z \in D$, then $z \in G$ and $qz \in G$. As z runs through D, the point qz runs through the rectangle D_q obtained from D by dilation by a factor of $|q|$ and rotation about the origin through the angle $\varphi_0 = \arg q$. Assume that $|q| \leq 1$ and $\varphi_0 \neq 0$. Let $\varphi_0 \neq \pi$. Since the domain of convergence of a Dirichlet series is convex, the domain G contains the convex hull of the rectangles D and D_q. The convex hull contains a disk $|z| < R$ with radius R depending on q, h_1, and ε, and this radius decreases to some limit value $R_0 > 0$ as $\varepsilon \to 0$. We see that the domain G contains the disk E_0: $|z| < R_0$. But then $G_q \supset E_0$. Hence (15) converges and represents the analytic function $\Phi(z) = 1/(z-\varepsilon)$ in E_0. But this cannot be, if ε is small. The case $\varphi_0 = \pi$ is even simpler to bring to a contradiction. Accordingly, a rectangle of small width and not symmetric with respect to the point $z = 0$ cannot serve as a domain D in which any analytic function can be expanded in a series (15).

BIBLIOGRAPHY

1. A. F. Leont′ev, *Exponential series*, "Nauka", Moscow, 1976. (Russian)

Translated by H. H. MCFADEN

Amer. Math. Soc. Transl.
(2) Vol. **140**, 1988

Remarks on Algebraic Supermanifolds

YU. I. MANIN

1. In a significant paper [1] E. Witten gave a connection between the problem of classifying vector bundles on open subsets of a special algebraic supermanifold and the problem of describing solutions of the supersymmetric Yang–Mills equation. This differential equation is under active discussion in contemporary theoretical physics. In two papers ([2] and [3]) G. M. Henkin (Хенкин) and I explained the cohomological nature of Witten's construction; however, in those papers we did not attempt to go beyond the limits of usual algebraic and analytic spaces with nilpotents. The goal of the present paper is to introduce the basic concepts of superalgebraic geometry, which allows us to describe Witten's ideas more precisely.

2. We shall use the elements of supercommutative algebra as presented, for example, in [4]. We recall what the mnemonic "super" means: all objects are equipped with a Z_2-grading, and the axioms of the fundamental structures differ from their usual versions by the factors ± 1 that take this grading into account. Thus, the ring $A = A_0 + A_1$ (where $\{0,1\} \in Z_2$) is supercommutative if $ab = (-1)^{\overline{a}\overline{b}} ba$; in formulas of this type it is tacitly assumed that $a, b \in A$ are homogeneous elements of degrees $\overline{a}$ and $\overline{b}$, respectively. We assume throughout this paper that 2 is invertible; then A_1 consists of nilpotents.

As in commutative algebra, every left A-module M has a standard structure of a right A-module: $ma = (-1)^{\overline{a}\,\overline{m}} am$. This allows us to define a tensor product. The usual isomorphisms of multilinear algebra are also equipped with signs: $M_1 \otimes M_2$ is identified with $M_2 \otimes M_1$ by the mapping $m \otimes n \mapsto (-1)^{\overline{m}\,\overline{n}} n \otimes m$. The symmetric algebra $S_A(M)$ is defined as the universal object for the mapping of M into supercommutative A-algebras. The change-of-parity functor Π is defined

1980 *Mathematics Subject Classification* (1985 *Revision*). Primary 14A20, 32L25, 58A50; Secondary 81E13.

Translation of Algebra (Collection in Honor of the Ninetieth Anniversary of the Birth of O. J. Schmidt; A. I. Kostrikin, editor), Izdat. Moskov. Gos. Univ., Moscow, 1982, pp. 95–101; MR **86i**: 58011.

by $(\Pi M)_0 = M_1$, $(\Pi M)_1 = M_0$, and the right multiplications on M and ΠM coincide. A free A-module of rank $r|s$ is a module isomorphic to $A^r \oplus (\Pi A)^s$. It is often convenient to call the modules A^r "purely even", and the $(\Pi A)^s$ "purely odd".

3. DEFINITION. A *superscheme* is a pair $(X, \mathscr{O})$ consisting of a topological space and a sheaf of supercommutative rings $\mathscr{O} = \mathscr{O}_0 + \mathscr{O}_1$ on it such that $(X, \mathscr{O}_0)$ is a scheme and $\mathscr{O}_1$ is a coherent sheaf of $\mathscr{O}_0$-modules.

Analytic superspaces are defined analogously.

4. EXAMPLES. a) Let $(X, \mathscr{O}_X)$ be a scheme in the usual sense, and let E be a locally free sheaf of $\mathscr{O}_X$-modules. We make X into a ringed space with the sheaf of Grassmann algebras $\Lambda(E) = \Lambda_0 + \Lambda_1$, where $\Lambda_0 = \sum_{i\geq 0} \Lambda^{2i}(E)$ and $\Lambda_1 = \sum_{i\geq 0} \Lambda^{2i+1}(E)$. Then $(X, \Lambda(E))$ is a superscheme. If X is reduced, then the ideal $E\Lambda(E)$ coincides with the nilradical $N \subset \Lambda(E)$, and $\mathscr{O}_X$ and E can be recovered from $\Lambda(E)$ by the formulas $\mathscr{O}_X = \Lambda(E)/N$ and $E = N/N^2$.

b) Generalizing the last remark, for any superscheme $(X, \mathscr{O})$ we consider the nilradical $N \subset \mathscr{O}$ and construct the associated graded object

$$\operatorname{Gr} X = \left(X, \operatorname{Gr}\mathscr{O} = \bigoplus_{i>0} N^i/N^{i+1}\right).$$

The simplest class of superschemes is the class of smooth superschemes, for which $X_{\text{red}} = (X, \mathscr{O}/N)$ is a smooth scheme, N/N^2 is locally free over X_{red}, and $\operatorname{Gr}\mathscr{O} = \Lambda_{\mathscr{O}/N}(N/N^2)$. In particular, N is generated by the odd sections, so that it would be more correct to write the last formula in the form

$$\operatorname{Gr}\mathscr{O} = \Lambda_{\mathscr{O}/N}(\Pi(N/N^2)) = S_{\mathscr{O}/N}(N/N^2).$$

In the C^∞ category, smooth superschemes correspond to supermanifolds, with local coordinate systems $(x^\alpha, \theta^\alpha)$, with commuting and anticommuting coordinates, respectively, which are studied in detail in [4]. It is known that in the C^∞ category any supermanifold X is isomorphic to $\operatorname{Gr} X$; that is, its structure sheaf is the Grassmann algebra of the sheaf of sections of some vector bundle over a differentiable manifold. Using the technique of nilpotent extensions, a technique which is standard in algebraic geometry, we shall show that this is not true in the algebraic and analytic categories.

5. LEMMA. *Let $(X, \mathscr{O}_X)$ be a smooth k-scheme. Then there is a one-to-one correspondence between the following sets:*

a) *sheaves of commutative rings $\tilde{\mathscr{O}}_X$ on X, included in an exact sequence of the form $0 \to \mathscr{J} \to \tilde{\mathscr{O}}_X \to \mathscr{O}_X \to 0$ (where $\mathscr{J}^2 = 0$ in $\tilde{\mathscr{O}}_X$ and $\mathscr{J}$ as an $\mathscr{O}_X$-module isomorphic over $\mathscr{O}_X$)*, that are locally split, i.e.,*

$$(\forall x \in X \exists \mathscr{U} \ni x, \tilde{\mathscr{O}}_X \simeq \mathscr{O}_X \oplus \mathscr{J} \text{ on } \mathscr{U});$$

*Editor's note. There is an apparent omission in this parenthetic statement in the original.

b) *cohomology classes in* $H^1(X, TX \otimes I)$ *up to the action of the group* $\operatorname{Aut} I$, *where* TX *is the tangent sheaf to* X.

For the proof see, for example, [5], where many more general facts are proved. Here we note only that the cohomlology class $h \in H^1(X, TX \otimes I)$ corresponding to the extension $0 \to I \to \tilde{\mathscr{O}}_X \to \mathscr{O}_X \to 0$ is the class in $\operatorname{Ext}^1_{\mathscr{O}_X}(\Omega^1\mathscr{O}_X, I)$ of the extension of $\mathscr{O}_X$-modules

$$0 \to I \approx dI \to \Omega^1\tilde{\mathscr{O}}_X \to \Omega^1\mathscr{O}_X \to 0;$$

$\Omega^1\tilde{\mathscr{O}}_X$ is an $\mathscr{O}_X$-module, since $I\, dI = 0$. The Čech cocycle representing h on the covering where $\tilde{\mathscr{O}}_X$ is locally split is the obstruction to the existence of a global projection $\tilde{X} \to X$, where $\tilde{X} = (X, \mathscr{O}_X)$.

From this lemma it is not hard to obtain a classification of the supermanifolds of "odd dimension two" (odd dimension is the rank of N/N^2 in the notations of §4).

6. PROPOSITION. *Let* X *be a smooth* k*-scheme, and* E *a locally free sheaf of* $\mathscr{O}_X$*-modules of rank* 2. *Then the superschemes* $X_s = (X, \mathscr{O}_X)$ *for which* $\operatorname{Gr} X_s$ *is isomorphic to* $(X, \Lambda(E))$ *and* $\mathscr{O}_0$ *is locally split are classified by the elements of the set* $H^1(X, TX \otimes \Lambda^2 E)/\operatorname{Aut} E$.

PROOF. By the preceding lemma, the sheaves $\mathscr{O}_0$ are classified by the corresponding cohomology groups, and the inclusion of $\mathscr{O}_1$ is equivalent to giving an $\mathscr{O}_X$-linear isomorphism of $\Lambda^2(E)$ with the nilradical $\mathscr{O}_0$.

In an analogous way one could classify the supermanifolds of odd dimension 3 and 4.

7. EXAMPLE. Let $X = P^n$ and $\Lambda^2(E) \simeq \mathscr{O}(k)$. We realize P^n as the space of lines in a linear space V. Set $\tilde{V} = \mathscr{O}_{P^n} \otimes V$. Using the exact sequence, which goes back to Euler, of the form

$$0 \to \mathscr{O}(k) \to \tilde{V}(k+1) \to TP(k) \to 0$$

and Serre's theorem, we find that the cases when $H^1(TP^n \otimes \Lambda^2(E)) \neq 0$ are exhausted by the following:

$$n = 1, k \leq -4; \quad H^1 = V \otimes S^{|k|-3}(V)/S^{|k|-2}(V), \quad \dim H^1 = |k| - 3.$$

In order to classify the "projective superlines" $P^{1|2}$ completely, we also need to take into account a theorem of Grothendieck, according to which $E \simeq \mathscr{O}(m) \oplus \mathscr{O}(n)$, $m + n = k$, and to take the quotient modulo $\operatorname{Aut} E$. Finally, over an algebraically closed field we obtain the following list of superlines $P^{1|2}$ that are not isomorphic to $\operatorname{Gr} P^{1|2}$:

$$P_h^{1|2}(m,n); \quad m+n \leq -4, \quad h \in P(H^1(P^1, \mathscr{O}_{P^1}(m+n+2))) = P^{|m+n|-4},$$
$$\operatorname{Gr} P_h^{1|2}(m,n) = (P^1, \Lambda(\mathscr{O}(m) \oplus \mathscr{O}(n))).$$

The case $m+n = -4$ is interesting, for it corresponds to a "rigid" series of nonvarying superlines. One ought to calculate the supergroups of automorphisms of this series and their action on the nontrivial cohomology spaces—can they realize some representations constructed independently in the theory of Lie superalgebras?

8. Picard group. We denote by $\operatorname{Pic} X$ the group of classes of even invertible sheaves on X. (In order to obtain the full Picard group, we need to add the second-order class of the sheaf $\Pi\mathscr{O}_X$; it generates a subgroup Z_2 which is a direct summand.) Since on a covering that locally trivializes a sheaf the transition functions are purely even invertible sections of $\mathscr{O}_X$, we have, as in the classical case, $\operatorname{Pic} X = H^1(X, \mathscr{O}_0^*)$.

Every homogeneous section of an invertible sheaf defines the closed supersubscheme of zeros of this section: under the local identification of the sheaf with $\mathscr{O}_X$, a section generates an ideal that is independent of the choice of identification. The zeros of an even section may be considered as super analogues of the usual Cartier divisors, having codimension $(1\,|\,0)$. The zeros of an odd section define a "divisor of codimension $(0\,|\,1)$", the support of which is the entire superscheme. Conversely, every Cartier superdivisor, which by definition is given by an invertible ideal $\mathscr{J} \subset \mathscr{O}$, defines an invertible sheaf $\mathscr{J}^{-1}$, the zeros of whose sections form this superdivisor.

9. Standard projective superspace. Let V be a linear superspace of dimension $(m+1\,|\,n)$ over a field k. We define the standard projective superspace $P^{m|n}$ by $\operatorname{Proj} S_k(V^{m+1|n})$. Setting $V_{0,1} = (V^{m+1|n})_{0,1}$, $P^m = \operatorname{Proj} S_k(V_0)$, and $\tilde{V}_1 = \mathscr{O}_{P^m} \otimes V_1$, we can identify $P^{m|n}$ with $(P^m, \Lambda(\tilde{V}_1(-1)))$ in the spirit of our earlier discussion.

It is not hard to see that

$$\mathscr{O}^*_{P^{m|n},0} = \mathscr{O}^*_{P^m} \times \Big(1 + \sum_{i\geq 0} \Lambda^{2i}(\tilde{V}_1)(-2i)\Big).$$

From this it follows that

$$\operatorname{Pic} P^{m|n} = \begin{cases} \operatorname{Pic} P^m = Z, & \text{if } m > 1, \\ \operatorname{Pic} P^1 \times H^1(P^1, \Lambda^2(\tilde{V}_1)(-2)) & \\ \quad = Z \times \Lambda^2 V_1, & \text{if } m = 1. \end{cases}$$

The subgroup Z is generated by the standard sheaf $\mathscr{O}(1)$.

10. Witten manifold. We consider two dual linear superspaces $V = V^{m+1|n}$ and $\hat{V} = \hat{V}^{m+1|n}$, related by the nondegenerate even scalar product $V \otimes \hat{V} \to k$. On the product $P(V) \times P(\hat{V})$ there is an invertible sheaf $\mathscr{O}(1,1)$ with the space of sections $(V \otimes \hat{V})^* = \mathscr{L}(V \otimes \hat{V}, k)$. The pairing between V and $\hat{V}$ corresponds to a canonical section s of this sheaf. We denote by $L = L_n$ the supermanifold of zeros of this section. Obviously, $L_{\text{red}} = P^m \times \hat{P}^m$ can be identified with the flag manifold (point in $P^m \subset$ hyperplane in P^m). We denote by $L^{(n)}$ the nth infinitesimal neighborhood of L_{red} in $P^m \times \hat{P}^m$, given by the ideal $\mathscr{J}^{n+1}$, where $\mathscr{J}$ is the ideal of L_{red}. Then L_n is a superextension of $L^{(n)}$: there is a canonical surjective mapping $L_n \to L^{(n)}$ which is the identity on L_{red}.

For this construction it is simplest to work with the $Z \times Z$ graded rings whose projective spectra are L_n and $L^{(n)}$, respectively. Let (z_i, θ_α) be a basis of V^*, where the z_i are the even generators and the θ_α are the odd generators, and let (z^i, θ^α) be the dual basis in $\hat{V} = V^*$. Then

$$L^{(n)} = \operatorname{Proj} k[z_i, z^j]/(z_i z^i)^{n+1},$$
$$L_n = \operatorname{Proj} k[z_i, \theta_\alpha; z^j, \theta^\beta](z_i z^i + \theta_\alpha \theta^\alpha), \quad \alpha = 1, \dots, n.$$

Since $(\theta_\alpha \theta^\alpha)^{n+1} = 0$, we have

$$(z_i z^i)^{n+1} = [(z_i z^i + \theta_\alpha \theta^\alpha) - \theta_\alpha \theta^\alpha]^{n+1} = w(z_i z^i + \theta_\alpha \theta^\alpha)$$

for a suitable w. Consequently, the obvious imbedding $k[z] \hookrightarrow k[z, \theta]$ induces a homomorphism of quotient rings; it is not hard to see that its kernel is trivial.

11. Connection with the Yang–Mills equations. In applications there is particular interest in the case $m = 3$, $k = C$. We set $T = (V^{4|n})_0 = C^4$. Then $L_{\text{red}} = F(1, 3; T)$ is the space of $(1, 3)$-flags in T. The Grassmannian $\operatorname{Gr}(2, T)$ is a complex compact model of Minkowski space; in particular, it is equipped with a canonical complex conformal metric: see, for example, [3]. We choose a connected, simply connected subset $\mathscr{U} \subset \operatorname{Gr}(2, T)$, whose intersection with any line in $\operatorname{Gr}(2, T) \subset P(\Lambda^2 T)$ is also connected and simply connected. We set $L_{\text{red}}(\mathscr{U}) = \{\ell \in F(1, 3; T) \colon \exists x \in \mathscr{U}$ such that the plane in T corresponding to x contains the 1-component and is contained in the 3-component of the flag $\ell\}$. Analogously we define $L_n(\mathscr{U})$ and $L^{(n)}(\mathscr{U})$, induced by the ringed space structures on L_n and $L^{(n)}$, respectively. For each point $x \in \mathscr{U}$ the manifold $L_{\text{red}}(x)$ is the quadric $CP^1 \times CP^1$. We say that a vector bundle E over $L(\mathscr{U})$ or $L^{(n)}_{\text{red}}(\mathscr{U})$ is *$\mathscr{U}$-trivial* if its restrictions to all the $L(x)$, $x \in \mathscr{U}$, are trivial.

In [2] and [3] it is shown that the category of $\mathscr{U}$-trivial vector bundles on $L_{\text{red}}(\mathscr{U})$ is equivalent to the category of Yang–Mills fields on $\mathscr{U}$, and the scalar and spinor fields and equations on them turn into cohomology classes on $\mathscr{U}$ and conditions for their prolongation into different neighborhoods $L^{(i)}(\mathscr{U})$, $i \leq 3$. Replacement of $L^{(3)}(\mathscr{U})$ by $L_3(\mathscr{U})$, proposed by Witten, allows us to get rid of the separate consideration of cohomology classes, interpreting them as classifiers of prolongations or obstructions to prolongations of a single superbundle. The details will be given elsewhere.

REFERENCES

1. Edward Witten, *An interpretation of classical Yang–Mills theory*, Phys. Lett. **77B** (1978), 394–398.

2. G. M. Henkin and Yu. I. Manin, *Twistor description of classical Yang–Mills–Dirac fields*, Phys. Lett. **95B** (1980), 405–408.

3. ——, *On the cohomology of twistor flag spaces*, Compositio Math. **44** (1981), 103–111.

4. D. A. Leĭtes, *Introduction to the theory of supermanifolds*, Uspekhi Mat. Nauk **35** (1980), no. 1 (211), 3–57; English transl. in Russian Math. Surveys **35** (1980).

Translated by J. S. JOEL

Amer. Math. Soc. Transl.
(2) Vol. **140**, 1988

On Some Connections Between Vector Optimization and Geometric Optimization and Their Economic Applications

K.-H. ELSTER AND R. HEINE

§1. Introduction and basic concepts

The problems which arose in the practical domains of technology, economics, etc., whose solutions led to the origins of vector optimization were often extremal problems in which it was required to find an extreme value of a function (objective function, utility function) under given constraints. Because of this there was, on the one hand, the advantage of a precisely formulated mathematical problem that made possible the formulation of theorems and the development of numerical methods. On the other hand, the formulation was often inadequate because it did not reflect the real relationships completely, for in practical applications the comparison of different solutions was often impossible with the use of a single criterion or could only be accomplished with significant simplifications.

When several criteria are taken into consideration, one enters the domain of *multiple criteria decision making* or *multicriteria analysis*, which allows a more realistic expression of the actual mathematical relationships involved, but which introduces difficulties into the mathematics. This leads to the concept of "the solutions" of a given problem. When there are several criteria, it is necessary to realize that in general these criteria conflict with each other, in the sense that the admissible point x may be better than another admissible point y with respect to one criterion but worse than y with respect to another.

If we write DM (decision maker) for a solution-finding algorithm, then two ways of proceeding might be considered, each of them clearly inappropriate in the kinds of situations we have been discussing:

(a) DM may choose a solution without taking any constraints into consideration;

1980 *Mathematics Subject Classification* (1985 *Revision*). Primary 90A05, 90C31.
Translation of Wiss. Z. Tech. Hochschule Ilmenau **30** (1984), no. 5, 43–54; MR **86b**:90119.

(b) DM may be given in advance constraints under which the problem can be solved in the usual form, i.e., with the use of a unique criterion.

The methods used in vector optimization are found to lie between these two boundary situations and are based on models which are partially defined by necessary mathematical assumptions and partially by information provided by the DM to the model or the system.

Here one encounters interesting and important practical and theoretical aspects, for in the usual (scalar) optimization the DM completely changes the mathematical model, while in multicriteria problems the DM actively participates in the construction of a solution (this is especially clear in interactive methods).

We will not dwell on the analysis of the concept "multicriteria optimization problems" any further here. We agree to understand by this term the collection of practical and theoretical results that exist in connection with problems of finding a solution or with problems of optimization when many objective functions are involved.

At the present time mathematical theories in the discipline of mathematical optimization seem to be developing as a result of the analysis of particular applications. In the theory of vector optimization itself, we will again distinguish between numerical solution methods and applications. Information about the rapid development of this domain can be obtained by consulting two bibliographies on vector optimization ([1] and [9]).

In our opinion a comparison of the publications in this area allows one to notice the following important fact. In socialist countries today there are many works on the theory and engineering-technical applications of vector optimization. These are, however, not concerned with the economic applications of this discipline.

We feel that the possibility of using vector optimization for the solution of economic problems has not been exhausted by any means. It would be very desirable to obtain a change in this situation, and valuable to accomplish this from the start with the aid of an international exchange of experience between mathematicians working in the area, independent of whether they are carrying out theoretical or practical investigations.

In this paper we will introduce several basic concepts of vector optimization and then give examples from one class of special problems of vector optimization, namely geometric problems. We will also present some new results of vector-signomial optimization.

If the problem is to select a solution, then there must be understood a set G of admissible solutions (operations, controls) and DM must:

(a) choose a single solution of G as "best"; or

(b) choose a subset of G as a set of "good" solutions; or

(c) order the solutions of G from best to worst.

Here the finite or infinite set G can be given either explicitly (for instance as a list) or inexplicitly (for instance, in the form of a system of inequalities). This leads to a classification of the problems of selecting a solution in the presence of many criteria (*multiple criteria decision making*—MCDM).

If the given set G consists of a finite number of explicitly given alternatives from which DM must make a selection, then we speak of *multiple attribute decision making* (MADM). These problems clearly have a great practical value, and thus special approaches have been developed for them. It is hardly possible to give an indication of the numerous applications of these problems.

If the given set G consists of an infinite number of inexplicitly given alternatives (operations, controls), then we speak of *multiple objective decision making* (MODM). From now on we will consider such problems only.

We consider the following problem of vector optimization:

$$\mathrm{MIN}\{g^z(t) | t \in G\}, \qquad \text{(VOP)}$$

where $G := \{t \in R^m | g^r(t) \leq 0\}, g^z \colon R^n \to R^{s+1}$, and $g^r : R^n \to R^p$; we denote by R^{s+1}_+ and R^p_+ the corresponding dominating cones.

A point $t^0 \in G$ is called a *P-minimal solution* of (VOP) if

$$t^0 \in \{t \in G | [g^z(G) + R^{s+1}_+ - g^z(t)] \cap [-R^{s+1}_+] = \{0\}\} =: G_{\mathrm{MIN}},$$

a P_{w}*-minimal solution* of (VOP) if

$$t^0 \in \{t \in G | [g^z(G) + R^{s+1}_+ - g^z(t)] \cap [\mathrm{int}(-R^{s+1}_+)] = \varnothing\} =: G_{\mathrm{wMIN}},$$

and a P_{e}*-minimal solution* of (VOP) if

$$t^0 \in \{t \in G | T(g^z(G) + R^{s+1}_+, g^z(t)) \cap [-R^{s+1}_+] = \{0\}\} =: G_{\mathrm{eMIN}}.$$

Here $T(A, a)$ is a tangent cone to the set $A = g^z(G) + R^{s+1}_+$ at the point $a = g^z(t)$.

The important problem arises of constructing suitable dual problems to the problem (VOP). We restrict ourselves here to an example of one possible approach.

For $\alpha \in \mathrm{int}\, R^{s+1}_+$ the problem (VOP) can be put in correspondence with the following problem:

$$\mathrm{MAX}\{L(U, t) | (U, t) \in G^*_\alpha\} \qquad (\mathrm{DOP}_\alpha)$$

where

$$G^*_\alpha = \{(U, t) \in R^{(s+1)\times p}_+ \times R^m | \alpha^{\mathrm{T}} \nabla_t L(U, t) = 0\}.$$

Here $\nabla_t L(U, t)$ denotes the Jacobian matrix (with respect to the variable t) of the function

$$L(U, t) := g^z(t) + U g^r(t), \qquad U \in R^{(s+1)\times p}_+, \quad t \in R^m.$$

There is also a possible approach in which (VOP) is "scalarized" and then the usual dual problem is constructed for the one-criterion problem thus obtained. By using a duality theorem for nonlinear programming, it is possible to obtain assertions indirectly for (VOP). This approach will be presented for a particular problem of geometric programming. The connection with the dual problem in the form (DOP_α) will be given in later investigations.

§2. Geometric programming and its economic applications

In recent years many papers have been published on the theory, numerical methods, and applications of geometric programming. Examples from engineering and technology provided the main impetus for the rapid development of this branch of the discipline of mathematical programming (see [5], [13], [2], and [4]). It turned out, however, that problems often arose in economics which led to geometric programming problems (see [8]); for instance,

a) models with production functions, for instance Cobb–Douglas functions;

b) production models with probabilistic constraints;

c) planning of capacities (of production levels);

d) marketing-mix problems (marketing studies);

e) investment planning;

f) distribution planning;

g) environmental problems.

The special structure of these problems is characterized by the fact that their objective functions and their constraint functions are so-called *posinomials* or *signomials*. Posinomial and signomial programming were developed for such functions (see [5], [11], [3], and [7]).

By using the special structure of these problems, it is possible to construct with the aid of geometric inequalities (this is the reason for the name "geometric programming") dual problems that are characterized by the fact that they have only linear constraint functions.

If a dual solution exists, one can find a solution to the original problem in the usual way. The advantage (theoretical and practical) of this approach consists in the fact that a solution of the original nonlinear problem is obtained by solving a system of linear equalities (in the case of posinomial programming) or a system of linear inequalities (in the case of signomial programming).

It turns out that models with a single objective function lead to an incorrect description for many practical problems. Thus it is important to simultaneously take into account several objectives in models of this kind. This leads to the consideration of multicriteria problems of geometric programming. At present some results on posinomial vector optimization are known [10]. In this paper we will present new results about multicriteria problems of signomial programming.

§3. Multicriteria problems of signomial programming

The multicriteria problem of signomial programming

$$(\mathrm{VOP})\quad \mathrm{MIN}\{g(t) \mid t \in G\},$$

$$G := \{t \in R^m \mid t > 0,\ 0 < (r_k g_k(t)^r) \le 1,\ k \in J^0\} \qquad (3.1)$$

with

$$g : R^m \to R^{s+1},\ s \geq 1; \qquad g(t) := (g_0(t), \dots, g_s(t))^{\mathrm{T}}, \tag{3.2}$$

where

$$g_k(t) := \sum_{i \in g[k]} s_i c_i \prod_{j=1}^{m} t_j^{a_{ij}}, \quad k \in J_{s+p}^0, \tag{3.3}$$

with

$$\left.\begin{array}{l} c_i > 0 \\ a_{ij} \in R \\ s_i \in \{-1, 1\} \end{array}\right\} \quad i \in g[k],\ k \in J_{s+p}^0,\ j \in J_m, \tag{3.4}$$

and

$$J_{s+p}^0 := \{\underbrace{0, 1, \dots, s}_{J_s^0}, \underbrace{s+1, \dots, s+p}_{J^0}\}, \qquad J_m := \{1, \dots, m\}, \tag{3.5}$$

$$g[k] := \{m_k, \dots, n_k\}, \qquad m_0 := 1,\ m_1 := n_0 + 1,\ m_{s+p} := n_{s+p-1} + 1, \quad n_{s+p} := n, \tag{3.6}$$

$$r_k \in \{-1, 1\}, \tag{3.7}$$

can be put in correspondence with the following one-criterion problem:

$$P_\alpha: \quad \min\{z_\alpha(t) | t \in G\} \tag{3.8}$$

with

$$z_\alpha(t) := \sum_{k=0}^{s} \alpha_k g_k(t), \qquad \alpha := (\alpha_0, \dots, \alpha_s)^{\mathrm{T}} > 0, \qquad \sum_{k=0}^{s} \alpha_k = 1. \tag{3.9}$$

Here g_k is as in (3.3), and, since z_α is a sum of signomials, it is a real objective function, so that P_α is a classical problem of signomial programming. This means that the theory of signomial programming in a modified form can be applied to problem P_α.

Thus, using this theorem, we place in correspondence with P_α the following dual problem:

$$P_\alpha^*: \quad \max\{v(\alpha, y, r_\alpha) | (\alpha, y, r_\alpha) \in G^*\}$$

$$G^* := \left\{ (\alpha, y, r_\alpha) \in R^{s+1} \times R^n \times R \,\middle|\, \alpha > 0,\ \sum_{k=0}^{s} \alpha_k = 1,\ r_\alpha \in \{-1, 1\}; \right.$$
$$\left. y \geq 0,\ r_k \lambda_k(y) \geq 0,\ k \in J_{s+p}^0,\ r_\alpha \sum_{k=0}^{s} \lambda_k(y) = 1,\ A^T y^s = 0 \right\} \tag{3.10}$$

with

$$v(\alpha, y, r_\alpha) := r_\alpha \left[\prod_{k=0}^{s} \prod_{i \in g[k]} \left(\frac{c_i}{y_i}\right)^{s_i y_i} \alpha_k^{\lambda_k(y)} \right.$$
$$\left. \cdot \prod_{k=s+1}^{s+p} \prod_{i \in g[k]} \left(\frac{c_i}{y_i}\right)^{s_i y_i} (r_k \lambda_k(y))^{\lambda_k(y)} \right]^{r_\alpha} \tag{3.11}$$

where

$$\lambda_k(y) := \sum_{i\in g[k]} s_i y_i, \quad k \in J^0_{s+p}. \tag{3.12}$$

Here c_i and s_i are as in (3.4), while r_k is as in (3.7). Also $y^s := (s_1y_1, \ldots, s_ny_n)^{\mathrm{T}}$. The matrix $A = (a_{ij})$ consists of the exponents a_{ij} of the function g_k.

REMARK. If $s_i = 1$, $i \in g[k]$, $k \in J^0_{s+p}$, and $r_k = 1$, $k \in J^0$, then the problem becomes a multicriteria problem of posinomial programming.

We assume the following hold for problem P^*_α:

(i) $r(A) < n$ ($r(A)$ is the rank of the matrix A); and

(ii)
$$y_i = 0 \ \forall i \in \bigcup_{k=0}^{s} g[k] \Leftrightarrow \sum_{k=0}^{s} \lambda_k(y) = 0;$$
$$y_i = 0 \ \forall i \in g[k] \Leftrightarrow \lambda_k(y) = 0, \ k \in J^0.$$

Assumption (i) guarantees the existence of a solution to the system of inequalities (a).

If $r(A) = n - 1$, then the degree of difficulty, defined as $D := n - r(A) - 1$, is equal to zero, and consequently problem P^*_α has a unique solution. If $D > 0$, there are several distinct solutions, out of which the appropriate method chooses optimal ones.

Assumption (ii) is necessary because of continuity considerations.

If $y_i = 0$ and $\lambda_k(y) = 0$, $k \in J^0$, then we set

$$y_i^{-y_i} := 1, \qquad (\lambda_k(y))^{\lambda_k(y)} := 1.$$

Since in problem P_α the objective function z_α and the set of admissible solutions G are nonconvex, the Kuhn–Tucker conditions are not satisfied.

Then we will speak of so-called *quasi-solutions* (Kuhn–Tucker points) and will write

$$z_\alpha(t^0) =: \operatorname*{qmin}_{t\in G} z_\alpha(t), \qquad \operatorname*{qmax}_{(\alpha,y,r_\alpha)\in G^*} V(\alpha, y, r_\alpha) := v(\alpha, y^0, r_\alpha)$$

where "qmin" is called a *quasi-minimum* and "qmax" a *quasi-maximum.* The vector t^0 denotes a quasi-solution of problem P_α, and the vector (α, y^0, r_α) a quasi-solution of problem P^*_α.

DEFINITION 1. Let $L\colon G \times R^p_+ \to R$, given by

$$L(t, \mu) := r_\alpha \ln r_\alpha z_\alpha(t) + \sum_{k=s+1}^{s+p} \mu_k r_k \ln r_k g_k(t),$$

where

$$0 \neq r_\alpha = \operatorname{sgn}\left(\min_{t\in G} z_\alpha(t)\right),$$

be the Lagrangian for problem P_α.

A vector t^0 such that

$$\text{(i) } t^0 \in G,$$

$$\text{(ii) } t_p^0 \left(\frac{\partial L}{\partial t_p}\right)_{t_p = t_p^0} = 0, \quad p \in J_m,$$

$$\text{(iii) } \mu_k^0 \ln r_k g_k(t^0) = 0, \quad k \in J^0,$$

is called a *quasi-solution* of problem P_α. We denote the set of all quasi-solutions of problem P_α by $(G)^0$.

It is possible to define quasi-solutions for the problem P_α^* in an analogous manner.

DEFINITION 2. Let $L^* \colon G^* \times R^{m+1} \times R_+^p \to R$, given by

$$L^*((\alpha, y, r_\alpha); \nu, \rho) := r_\alpha \ln r_\alpha v(\alpha, y, r_\alpha) + \nu_\alpha \left(r_\alpha \sum_{k=0}^{s} \lambda_k(y) - 1 \right) + \sum_{j=1}^{m} \nu_j \left(\sum_{k=0}^{s+p} \sum_{i \in g[k]} a_{ij} s_i y_i \right) + \sum_{k=s+1}^{s+p} \rho_k r_k \lambda_k(y),$$

be the Lagrangian for the problem P_α^*.

A vector (α, y, r_α) such that

$$\text{(i) } (\alpha, y, r_\alpha) \in G^*,$$

$$\text{(ii) } \left(\frac{\partial L^*}{\partial y_i}\right)_{y_i = y_i^0} = 0, \; i \in g[k], \; k \in J_{s+p}^0,$$

$$\text{(iii) } \rho_k^0 r_k \lambda_k(y^0) = 0, \; k \in J^0,$$

is called a *quasi-solution* of problem P_α^*. $(G^*)^0$ is the set of all quasi-solutions of problem P_α^*.

Using quasi-solutions, we can now define quasi-dual problems.

DEFINITION 3. Let problems P_α and P_α^* be given. Problem P_α^* is said to be *quasi-dual to* P_α if for every quasi-solution t^0 of P_α there exists a quasi-solution (α, y^0, r_α) of P_α^* such that

$$z_\alpha(t^0) = v(\alpha, y^0, r_\alpha).$$

If P_α is also quasi-dual to P_α^*, then the problems are said to be simply *quasi-dual.*

The following assertions can be proved about problems P_α and P_α^*.

THEOREM 1. *Problems P_α and P_α^* are quasi-dual.*

THEOREM 2. *Let* $\operatorname{int} G \neq 0$ *be satisfied for problem P_α. If the vector t^0 is a quasi-solution of P_α, then the following hold:*

(i) *there exists a quasi-solution (α, y^0, r_α) for problem P_α^*;*

(ii) $z_\alpha(t^0) = v(\alpha, y^0, r_\alpha)$;

$$\text{(iii)} \quad c_i \prod_{j=1}^{m} (t_j^0)^{a_{ij}} = \begin{cases} \dfrac{r_\alpha}{\alpha_k} y_i^0 v(\alpha, y^0, r_\alpha), & i \in g[k], \; k \in J_s^0; \\[2ex] \dfrac{y_i^0}{r_k \lambda_k(y^0)}, & i \in g[k], \; r_k \lambda_k(y^0) > 0, \; k \in J^0. \end{cases}$$

REMARK. It is possible to define the original variables by using property (iii). Theorem 2 follows directly from Theorem 1.

THEOREM 3. *Let* $\operatorname{int} G^* \neq 0$ *be satisfied for problem* P^*_α. *If the vector* (α, y^0, r_α) *is a quasi-solution of* P^*_α, *then the following assertions hold:*

(i) *there exists a quasi-solution* t^0 *for problem* P_α;

(ii) $z_\alpha(t^0) = v(\alpha, y^0, r_\alpha)$;

$$\text{(iii)}\quad c_i \prod_{j=1}^{m} (t_j^0)^{a_{ij}} = \begin{cases} \dfrac{r_\alpha}{\alpha_k} y_i^0 z_\alpha(t^0), & i \in g[k],\ k \in J_s^0; \\[2ex] \dfrac{y_i^0}{r_k \lambda_k(y^0)}, & i \in g[k],\ r_k\lambda_k(y^0) > 0,\ k \in J^0. \end{cases}$$

Theorem 3 is the analogue of Theorem 2 for problem P^*_α, and follows directly from Theorem 1.

If a signomial programming problem P_α has a solution, then it is possible to prove the following theorem with the aid of Theorem 2.

THEOREM 4. *Let problems* P_α *and* P^*_α *be given. If a vector* t^* *is an optimal solution of* P_α, *then the following relations hold:*

(i) $(G^*)^0 \neq \varnothing$;

$$\text{(ii)}\qquad z_\alpha(t^*) = \min_{t\in G} z_\alpha(t) = \min_{(\alpha,y,r_\alpha)\in(G^*)^0} v(\alpha, y, r_\alpha) := v(\alpha, y^*, r_\alpha).$$

Conversely, from the existence of quasi-solutions for P^*_α follows the existence of solutions of P_α.

THEOREM 5. *Let problems* P_α *and* P^*_α *be given. If there exists a vector* $(\alpha, y^*, r_\alpha) \in (G^*)^0$ *such that*

$$v(\alpha, y^*, r_\alpha) = \min_{(\alpha,y,r_\alpha)\in(G^*)^0} v(\alpha, y, r_\alpha) = \min z_\alpha(t),$$

then there exists a vector $t^* \in G$ *such that*

$$z_\alpha(t^*) = \min_{t\in G} z_\alpha(t) = v(\alpha, y^*, r_\alpha).$$

The following proposition shows the connection between solutions of a multicriteria problem (VOP) and the problem P_α.

THEOREM 6. *Let problems* (VOP) *and* P_α *be given. If the vector* t^0 *is a solution of* P_α, *then* t^0 *is a* P_e*-minimal solution of* (VOP).

§4. Example

We will present an application of this theorem in economics. We consider a simplified model in input-output form:

$$(\mathbf{E} - \mathbf{A})\mathbf{x} - \mathbf{y} \geq 0. \tag{4.1}$$

The elements a_{ij} of the $n \times n$ matrix $\mathbf{A}$ give the amount of the product produced in branch i of the system that is supplied by branch i to branch j for the production of a single item produced in branch j.

$\mathbf{y} := (y_1, \ldots, y_n)^{\mathrm{T}}$ is the vector of proper (final) products, the part of the total production that is no longer put back into the process of production.

$\mathbf{E}$ is the identity matrix.

$\mathbf{x}$ is the vector of total production (intermediate and finished products; i.e., the gross production).

Relation (4.1) shows that the amount of goods remaining after the series of productions may exceed the amount of proper products; the model is not concerned with what happens to these products.

The definition of the required Luptáčik resources uses the Cobb–Douglas production function

$$x_j = \varepsilon_j L_j^{\alpha_j} K_j^{\beta_j}, \quad j \in J_n. \tag{4.2}$$

The usage coefficient ε_j, $j \in J_n$, gives the change in the effectiveness of both the factors L_j and K_j, $j \in J_n$, used in the given period.

L_j, $j \in J_n$, characterizes the labor used.

K_j, $j \in J_n$, characterizes the cost of the materials required.

α_j and β_j, $j \in J_n$, are the product elasticity indices.

α_j gives the percent change in production in branch j if the labor force increases by 1% while the remaining factors remain constant.

β_j has an analogous meaning in terms of the materials required by the production process.

The system of inequalities (4.1) can be written in the form

$$\sum_{j=1}^{n} (\delta_{ij} - a_{ij}) x_j - y_i \geq 0, \quad i \in J_n, \tag{4.3}$$

where δ_{ij} is the usual Kronecker symbol.

$$\delta_{ij} = \begin{cases} 0 & \text{if } i \neq j, \\ 1 & \text{if } i = j. \end{cases}$$

Applying the function (4.2) for x_j, $j \in J_n$, gives

$$\sum_{j=1}^{n} (\delta_{ij} - a_{ij}) \varepsilon_j L_j^{\alpha_j} K_j^{\beta_j} - y_i \geq 0, \quad i \in J_n. \tag{4.4}$$

Multiplication of the ith inequality of (4.4) by

$$-\frac{1}{1 - a_{ii}} L_i^{-\alpha_i} K_i^{-\beta_i}, \quad i \in J_n,$$

and elementary transformations yield

$$\sum_{\substack{j=1 \\ j \neq i}}^{n} d_{ij} L_j^{\alpha_j} K_j^{\beta_j} L_i^{-\alpha_i} K_i^{-\beta_i} + \frac{y_i}{(1 - a_{ii})\varepsilon_i} L_i^{-\alpha_j} K_i^{-\beta_i} \leq 1, \tag{4.5}$$

where

$$d_{ij} := \frac{a_{ij}\varepsilon_j}{(1 - a_{ii})\varepsilon_i}, \quad i \neq j.$$

Relation (4.5) shows that the sum of the parts of the total production which are used by the various branches of the production system and the proper products must be less than or equal to one.

Since the resources in the given system are only available in a limited amount, we must have $K_j \leq \overline{K}_j$, $j \in J_n$, or

$$\overline{K}_j^{-1} K_j \leq 1, \quad j \in J_n. \tag{4.6}$$

$\overline{K}_j$, $j \in J_n$, denotes the capacity available in branch j.

Relation (4.6) shows that the part of the available capacity that occurs in branch j cannot be greater than one.

As criteria according to which the resources L_j and K_j, $j \in J_n$, can be combined, we introduce into the model the following objective functions:

$$f_0(L) := \sum_{j=1}^{n} L_j, \tag{4.7a}$$

$$f_1(K) := \sum_{j=1}^{n} K_j. \tag{4.7b}$$

Here $f_0(L)$ characterizes the total labor used, and $f_1(K)$ the total cost of the materials required.

The problem is to determine the minimum of the function $f = (f_0, f_1)^{\mathrm{T}}$ under the constraints (4.5) and (4.6) and the positivity constraints $L_j > 0$ and $K_j > 0$, $j \in J_n$.

The solution of this problem yields for a given proper product the optimal distribution of labor and materials in the separate branches and simultaneously finds the optimal total production for each branch.

In order to write this problem in the form of a (VOP) problem, we set

$$\begin{aligned} t_j &:= L_j, \quad j \in J_n, \\ t_{n+j} &:= K_j, \quad j \in J_n, \\ C_j^k &:= d_{ij}, \quad j \neq k, \ j, k \in J_n, \\ C_k^k &:= \frac{y_k}{(1 - a_{kk})\varepsilon_k}, \quad k \in J_n. \end{aligned}$$

We obtain in this way the following problem of vector optimization:

$$\mathrm{MIN}\{g(t) | t \in G\}, \tag{VOP}$$

where

$$G := \{t \in R^m | t > 0,\ 0 < (r_k g_{k'}(t))^{r_k} \leq 1,\ k \in J^0\},$$
$$g\colon R^m \to R^2,\ g(t) = (g_0(t), g_1(t))^{\mathrm{T}},\ m = 2n,$$
$$g_0(t) := \sum_{j=1}^{n} t_j, \qquad g_1(t) := \sum_{j=n+1}^{2n} t_j,$$
$$g_{k'}(t) := t_k^{-\alpha_k} t_{n+k}^{-\beta_k} \left(\sum_{\substack{j=1 \\ j \neq k}}^{n} C_j^k t_j^{\alpha_j} t_{n+j}^{\beta_j} + C_k^k \right),\ k \in J_n,\ k' \in \{2, \ldots, n+1\}.$$

We will not formulate problems P_α and P_α^* here.

§5. Perspective

To conclude this note we would like to mention several problems outstanding in the domain of multicriteria analysis.

1. Further development of the theory of vector optimization.
2. The development of geometric vector optimization.
3. Study of the structure of domination.
4. Development, testing, and comparison of interactive methods.
5. Systematic investigations of the application of vector optimization to economic problems.

BIBLIOGRAPHY

1. A. Achilles, K.-H. Elster, and R. Nehse, *Bibliographie zur Vektoroptimierung (Theorie und Anwendungen)*, Math. Operationsforsch. Statist. Ser. Optim. **10** (1979), 277–321.

2. M. J. Rijckaert, *Engineering applications of geometric programming*, Optimization and Design (Internat. Summer School, Louvain, 1971; M. Avriel et al., editors), Prentice-Hall, Englewood Cliffs, N.J., 1973, pp. 196–220.

3. M. Avriel and A. C. Williams, *Complementary geometric programming*, SIAM J. Appl. Math. **19** (1970), 125–141.

4. Charles S. Beightler and Donald T. Phillips, *Applied geometric programming*, Wiley, 1976.

5. Richard J. Duffin, Elmor L. Peterson, and Clarence Zener, *Geometric programming: theory and application*, Wiley, 1967.

6. Arthur M. Geoffrion, *Proper efficiency and the theory of vector maximization*, J. Math. Anal. Appl. **22** (1968), 618–630.

7. R. Heine and K. Petrie, *Bibliographie zur geometrischen Optimierung*, Math. Operationsforsch. Statist. Ser. Optim. **14** (1983), 467–489.

8. Mikuláš Luptáčik, *Nichtlineare Programmierung mit ökonomischen Anwendungen*, Athenäum, Königstein/Ts., 1981.

9. Reinhard Nehse, *Bibliographie zur Vektoroptimierung—Theorie und Anwendungen*. II, Math. Operationsforsch. Statist. Ser. Optim. **13** (1982), 593–625.

10. Luis D. Pascual and Adi Ben Israel, *Vector-valued criteria in geometric programming*, Oper. Res. **19** (1971), 98–104.

11. V. Passy and D. J. Wilde, *Generalized polynomial optimization*, SIAM J. Appl. Math. **15** (1967), 1344–1356.

12. J. Ponstein, *Infima of sets, proper efficiency and multi-objective optimization*, Report 88 (OR-8202), Econometric Inst., Univ. Groningen.

13. Clarence Zener, *Engineering design by geometric programming*, Wiley, 1971.

Translated by S. TOLEDO